P9-DFR-438

INTELLIGENCE UNLEASHED

CREATING LEGO® NXT™ ROBOTS WITH JAVA™

Brian Bagnall
VARIANT PRESS

LONGWOOD PUBLIC LIBRARY

Copyright © 2011 by Brian Bagnall

All rights reserved, including the right of reproduction in whole or in part in any form.

Designed by Hayden Sundmark
Original artwork by Calvin Innes
Edited by Sylvia Philipps
Cover by Hayden Sundmark
Printed in Canada

Library and Archives Canada Cataloguing in Publication

Bagnall, Brian, 1972-

Intelligence unleashed : creating LEGO NXT robots with Java written by Brian Bagnall.

Includes index.

ISBN 978-0-9868322-0-8

1. Robotics--Popular works. 2. LEGO toys. 3. Java (Computer program language). 4. Robots--Programming. 5. Robots--Design and construction. I. Title.

TJ211.15.B343 2011   629.8'92   C2011-904987-2

VARIANT PRESS
3404 Parkin Avenue
Winnipeg, Manitoba
R3R 2G1

LEGO, LEGO MINDSTORMS NXT, and the brick configuration are trademarks of the LEGO Group.

The Windows logo is a trademark of Microsoft.

The Macintosh logo is a trademark of Apple.

LEGO, Microsoft, and Apple do not sponsor, authorize or endorse this book.

CONTENTS

Acknowledgements

Thanks to Lawrie Griffiths, Professor Roger Glassey, Sven Köhler and Andy Shaw for their incredible work on leJOS NXJ. And a big thanks to LEGO® for nurturing the amateur robotics community.

- Brian Bagnall

Preface

What is technology? Stone Age tools were considered technology at one time. According to pioneering computer programmer Alan Kay, "technology is anything that wasn't around when you were born." For the LEGO Mindstorms robot kit, this statement applies to most readers. Inventor Danny Hillis views technology as "the name we have for stuff that doesn't work yet."

Ten years ago the leJOS developers set out to improve the technology of our robot navigation. We thought it would be simple to make robots navigate reliably across the floors of our homes. Little did we realize the awesome task ahead.

By June 2009, the leJOS developers had a number of navigation classes that used a compass, GPS, and even Monte Carlo localization. Unfortunately none of them worked together very well. The package was restricted to two-wheeled caster robots—walkers, Segways, and other robots were excluded.

We decided to overhaul the navigation package to prepare it for modern probabilistic robotics, different robot types, diverse methods of localization, and pathfinding. Over the period of a year, we sent over 2000 individual e-mails to each other. Sometimes we felt frustrated or impatient, but our determination won out and we eventually produced an architecture we are confident about.

The results of our work are described within this book. The goal of leJOS NXJ has always been to include cutting edge programming technology so that you can rapidly prototype your idea in an hour instead of a month. With leJOS NXJ we are closer to that goal than ever.

What's in the book?

This book is about robot projects. If you are looking for a book to learn Java and LEGO NXT programming, look to the predecessor of this book, Maximum LEGO NXT: Building Robots with Java Brains.

The projects in this book are numerous and diverse. Technology such as vision, RFID (radio frequency identification), wireless communications and GPS feature prominently in some projects.

Navigation is a key robotics concept. A third of the book is dedicated to navigation. The definition of navigation is to manage or direct the course of. There are several aspects of managing the course of a robot:

- Localization – Where am I?
- Map Making – Where have I been?
- Pathfinding – How do I get there?
- Mission Planning – Where am I going?

Before you can confront the last three concepts above, you need to answer the first question: Where am I? Much of this book will examine ways to answer that question.

All code samples and errata are available from the following website: www.variantpress.com

July 12, 2011

CHAPTER 1

leJOS NXJ

TOPICS IN THIS CHAPTER

- Introducing NXJ
- Installation
- Uploading and Running Code

The software package used in this book allows you to program your NXT brick using Java. The package, known as leJOS NXJ, allows you to do things you can't do with other development environments. For example, the mapping project in chapter 13 plots real-time data on your PC from the NXT. NXJ also contains a pre-made library of sophisticated routines for handling navigation and other common robotics tasks.

This chapter gives a brief introduction to the leJOS NXJ project—a Java environment complete with threads, arrays, floating point numbers, recursion, garbage collection, and total control of the NXT brick. We will also set up leJOS NXJ on your computer and upload the firmware to your NXT brick.

Introducing NXJ

NXJ allows you to program your robot in Java, but are there really any differences between the official Java language and leJOS NXJ? Not really. NXJ gives you almost everything you get from standard Java:

- a Java Virtual Machine to run your code
- classes to handle computing tasks
- tools to compile code

The NXJ software allows you to control motors, read sensors, and have fun with robots.

So what does leJOS mean? The word *lejos* means far in Spanish. The first two letters, LE, are short for LEGO. The letters JOS are capitalized because those letters stand for Java Operating System. Since le means 'the' in several languages, this would mean *the Java Operating System*. NXJ refers to the part of the package that is specifically for the NXT brick, such as lejos.nxt.Motor (which is covered in chapter 3). NXT is a registered trademark, so we use NXJ to signify a link with Java while still indicating NXT compatibility.

The leJOS JVM is written in C code. The developers attempted to program the JVM in a platform independent style so it can be ported to other hardware, and we are constantly trying to improve this aspect. So far it has only appeared on the Gameboy Advance, and the code base has been used to create nxtOSEK, an ANSI C/C++ platform for the NXT.

There are also tools on the PC side to compile and upload code to the leJOS JVM. leJOS is multiplatform, and these days that means Windows, Linux, and Macintosh. leJOS NXJ is available for each of these platforms, allowing you to develop NXJ code under your favorite operating system.

Let's install leJOS NXJ and see what it can do. The following section will install the firmware to your NXT brick, but don't worry, it is not permanent. You can easily go back to the standard LEGO firmware at any time by using the LEGO software. It is also impossible to damage your brick by changing the firmware.

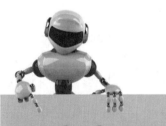

Just for fun!

The popularity of leJOS NXJ is increasing, and we regularly surpass the monthly download record of the previous leJOS RCX project. This probably indicates the NXT kit is more popular than the older Robotics Invention System.

Installing NXJ

The leJOS NXJ download contains the latest setup instructions. There is also an up-to-date tutorial on the leJOS website which describes these same procedures for Windows, Linux and Max OS X. If you experience any difficulty using the steps below, refer to the tutorial.

 Windows Preinstall

Before installing leJOS you will need to make sure you have a USB driver (you probably already do if you installed the LEGO software). You will also need to install the Java Developer Kit (minimum version 1.6).

1. You will need a suitable USB driver on your PC. If you have installed the LEGO Mindstorms software, a USB driver will already be installed. If you do not wish to install the LEGO software on your PC you can get a driver from the LEGO NXT download section at: mindstorms.lego.com/en-us/support/files/Driver.aspx

2. You will need at least version 1.6 of the Java Development Kit. Download and install the latest JDK from: www.oracle.com/technetwork/java

3. Environment variables indicate the locations of important files. If you want to be able to compile and run Java code from the command line, add the JDK bin directory to your PATH environment variable, so that commands such as `javac` and `java` can be called from a command prompt (this is not necessary for leJOS, however). Select Start > Control Panel > System. In Vista, click Advanced System Settings on the left side (Figure 1-1).

Just for fun!

We released NXJ in January 2007, and since then it has been downloaded over 200,000 times as of 2011. Germany is the top downloader, followed closely by the United States, then Great Britain, Canada, and France.

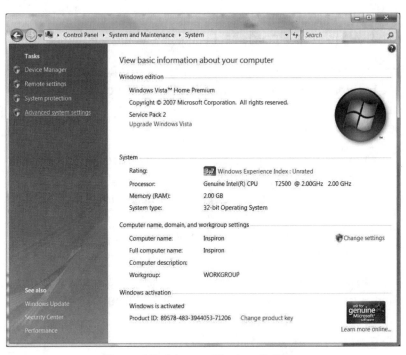

Figure 1-1: Advanced System Settings

4. Click the Advanced tab, then Environment Variables
   (Figure 1-2).

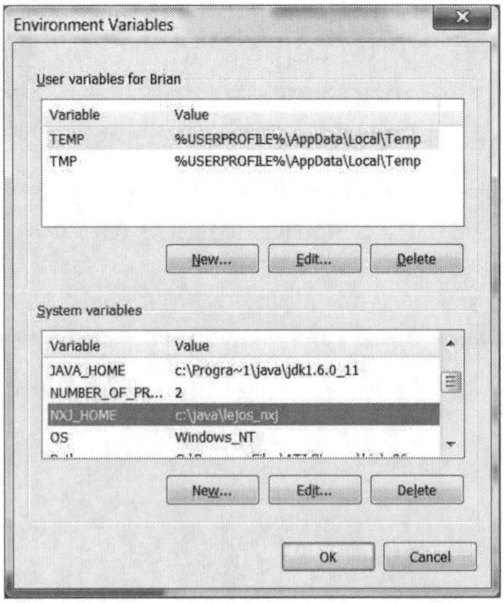

Figure 1-2: Environment Variables

5. Scroll down in your system variables and look for JAVA_HOME. If this is not set up, select New... For the Variable name, type JAVA_HOME. For the Variable value, type the location to your Java install.

6. To add Java to the path directory, select the Path variable and click Edit... (see Figure 1-3). Enter the following at the end of the existing Path and click OK when done:

   `;%java_home%\bin`

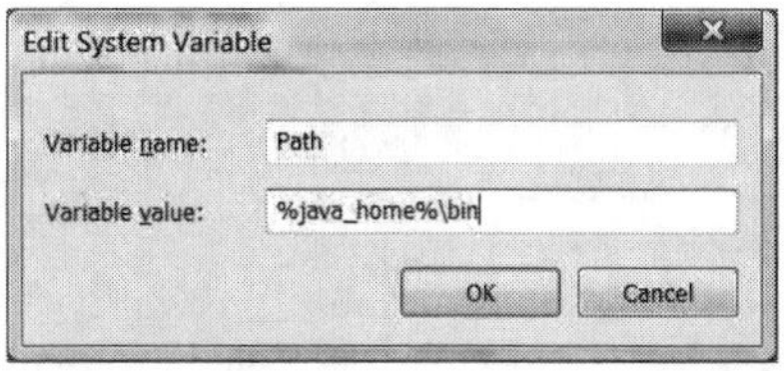

Figure 1-3: Java path system variable

NOTE: If you have multiple JDKs on your PC, and do not want to change JAVA_HOME, you can set LEJOS_NXT_JAVA_HOME instead.

Now let's test make sure Java is working. Launch a command prompt by selecting the Windows start menu. Type in CMD to run the command prompt. Type in 'java' to see if it has access to the Java binaries. If you get a help file listing it means it is working.

Troubleshooting:

Problem	Solution
'javac' is not recognized as an internal or external command, operable program or batch file.	1. Check PATH environment variable. Did you type it correctly? 2. Check JAVA_HOME. Is this the correct path?

 **Windows the Easy Way**

Now that you have Java installed and working, it's time to install NXJ. The easiest way to install it is to download the installer from www.lejos.org.

1. Download the Setup.exe file from the download section of www.lejos.org.

2. Run the install program and follow the instructions (Figure 1-4). The default directories should be fine.

---

leJOS NXJ Setup

Choose Destination Location

Where should leJOS NXJ be installed?

Setup will install leJOS NXJ in the following folder.

To install to this folder, click Next. To install to a different folder, click Browse and select another folder.

JDK version is 1.6.0_11

Destination Folder

C:\Program Files\leJOS NXJ   Browse...

InstallJammer

< Back   Next >   Cancel

Figure 1-4: Install wizard

---

3. When installation is complete, a program will start that will flash your NXT brick with new firmware. You can now continue on past the Linux and Mac installation sections.

 **Windows the Hard Way**

If you have already installed leJOS NXJ using the installer, you do not need to complete this section. The following steps will walk you through a manual install.

1. Your first step on the path to leJOS bliss is to download the latest version from www.lejos.org. This file ends with zip.

2. Unzip the contents into a directory. e.g. c:\java\lejos_nxj

3. Now we need to set some environment variables. Select Start > Control Panel > System and Maintenance > System (Figure 1-1).

4. Click the Advanced system settings, then Environment Variables (Figure 1-2).

5. Click New to create a new environment variable. It can either be in System variables (if multiple users will use leJOS) or User variables if your account is the only one using leJOS. Type NXJ_HOME as the variable name and add the leJOS directory (see Figure 1-5). Click OK.

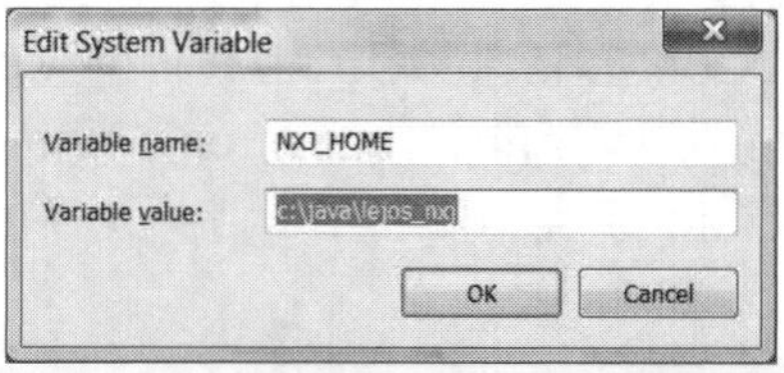

Figure 1-5: Setting the leJOS home directory

6. Finally, add the bin directory to your path so you can use the leJOS compiler tools from the command line. Add the following to the end of your path variable (see Figure 1-6).

   `;%NXJ_HOME%\bin`

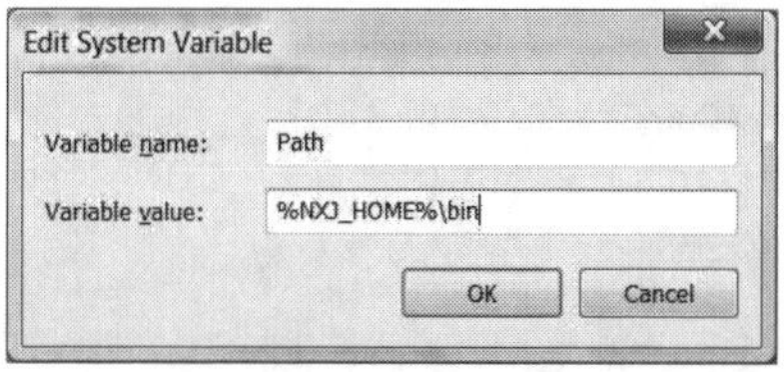

Figure 1-6: Setting the path to the leJOS binaries

That's all. Make sure you have already installed the LEGO NXT software, as its USB drivers are used. You can now skip down to Uploading Firmware.

Linux

Before getting started, download and install the latest version of the Java Development Kit (JDK) from Oracle. Your PATH must contain the JDK's bin directory. Also, make sure you have set the JAVA_HOME properly to the root directory of the JDK. If you have multiple versions of Java and would like to select one for leJOS, you can set LEJOS_NXT_JAVA_HOME instead.

USB Preparation

1. Install libusb so the leJOS tools can access your USB port. leJOS requires the legacy 0.1.12 release. You can find the library at: http://libusb.sourceforge.net. Ensure that the packages that leJOS NXJ are dependent on are on your system. To build the jlibnxt JNI library, which is used for USB access, you need the Development files for libusb (libusb-devel).

2. You will need to ensure that you have read and write access the NXT USB device in /dev/bus/usb. Use udev rules. How to do this may vary with different Linux systems.

To use udev rules, set up a file such as /etc/udev/rules.d/70-lego.rules and populate it with the following lines:

```
# Lego NXT brick in normal mode

SUBSYSTEM=="usb", DRIVER=="usb",
ATTRS{idVendor}=="0694",
ATTRS{idProduct}=="0002", GROUP="lego",
MODE="0660"

# Lego NXT brick in firmware update mode
(Atmel SAM-BA mode)

SUBSYSTEM=="usb", DRIVER=="usb",
ATTRS{idVendor}=="03eb",
ATTRS{idProduct}=="6124", GROUP="lego",
MODE="0660"
```

This relies on the username you are using being in the *lego* group. You can modify the file to your requirements. The two vendors are LEGO and Atmel. The rules are reloaded automatically. To activate the rules, turn off the NXT and turn it back on again.

leJOS NXJ Install

1. Download the tar.gz file from www.lejos.org.

2. Decompress the file into a directory, such as /opt/lejos_nxj/

3. Set your environment variable NXJ_HOME to the directory you installed leJOS.

4. Add the leJOS bin directory to your PATH. Depending on the privilege settings, you might need to adjust the execution permissions in the bin folder.

5. Your PATH must also contain the ant binary (ant 1.6 or above).

6. Now you need to build the distribution. Switch to the build folder and run ant. Note that depending on the privilege settings you might need to adjust the execution permissions in the release folder.

That's all. You can now skip down to Uploading Firmware.

 **Mac OS X**

Macintosh owners can download binary files compiled just for their system. These files are a universal build, meaning they will work on both Power PC and Intel based computers. A minimum version of Mac OS X 10.4 is required.

Make sure to download and install the Java Development Kit 1.6 or higher.

The leJOS tools for compiling and uploading Java code run in a shell environment, such as tcsh. Before you can do that, you will need to set up some environment variables for the tcsh shell.

1. Install the standard LEGO software, which will install the USB drivers.

2. Download the Mac OS X distribution from www.lejos.org.

3. Extract this file into a new location, such as /Applications/lejos_nxj.

4. If you use the administrator login, you will need to create (or edit) the file *.tcshrc* in your user home directory. Run textedit from your Applications folder.

5. Type the following two lines into the window (using the directory where you extracted leJOS), then save:

```
setenv NXJ_HOME /Applications/lejos_nxj
setenv PATH ${PATH}:${NXJ_HOME}/bin
```

6. Select your administrator directory (/users/administrator), and type *.tcshrc* as the file name. Uncheck the box saying "If no extension is provided, use '.txt'" before you save. Then you'll get a warning box suggesting these names are reserved for the system (see Figure 1-7). Click Use '.' and it will save.

NOTE: If you prefer to use csh instead of tcsh, you should instead edit/create the file *.cshrc* with the same lines.

7. Bring up a Terminal window and type *tcsh*. You'll now be in a tcsh shell.

8. Type *setenv* to make sure your PATH and NXJ_HOME variables are set up correctly. That's all! You are ready to test leJOS.

Just for fun!

Downloads of the Windows version are outpacing Linux and Macintosh combined by about three to one. 71% of downloads are Windows, followed by Linux at 14% and Macintosh at 10%.

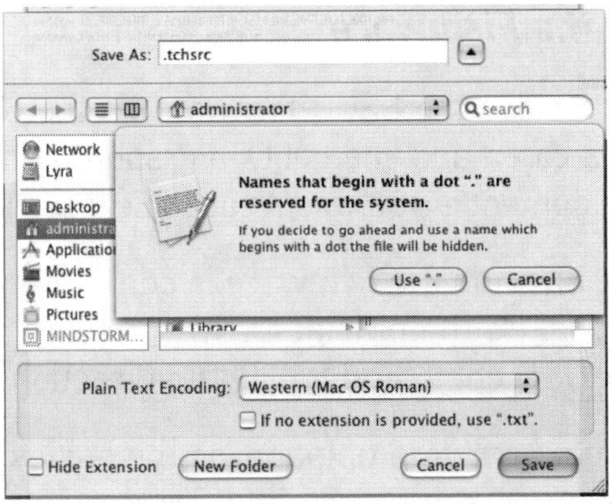

Figure 1-7: Setting the environment variables in Max OS X

Using leJOS NXJ

The USB cable is the only method for uploading firmware (Bluetooth cannot be used to flash firmware). LEGO even supplies a standard USB cable which is identical to a printer cable.

NOTE: When your NXT is plugged in using the USB cable, you can view it in the Device Manager for your operating system (see Figure 1-8).

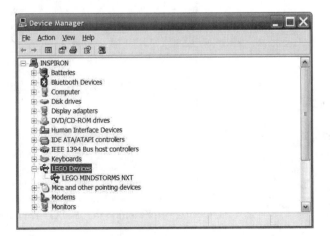

Figure 1-8: LEGO NXT in Devices

Uploading the Firmware

It is easy to replace the LEGO firmware on the NXT brick by using either a command line utility, a graphical utility, or from within a development environment called Eclipse (next chapter).

Don't worry, flashing the firmware is completely reversible. You can go back to the official LEGO firmware at any time. Plug in your USB cable and follow the directions.

1. A graphical firmware uploader is the easiest way to flash the brick (see Figure 1-9). To use this, run nxj-flashg from the bin directory and follow the instructions. Or, if you use the setup install, it will run automatically at the end of the software installation.

2. To flash using text tool, turn on your NXT brick, plug in your USB cable and type:

```
nxjflash
```

3. After a very brief moment you will see the leJOS NXJ logo and a menu system will appear. Your NXT brick is now ready to accept Java code.

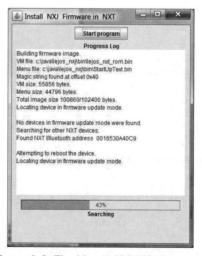

Just for fun!

The leJOS NXJ project contains over 200,000 lines of code.

Figure 1-9: Flashing leJOS NXJ firmware

Compiling and Uploading Java Code.

In this section you can try compiling and uploading some sample code from the command line. If you would rather use a graphical user interface, feel free to skip this section and jump straight to Chapter 2. Windows users can enter the command line by selecting Start > Run and then typing cmd (Click OK).

1. From a command line prompt, change to the samples\tune directory where you installed leJOS.

```
cd \java\lejos_nxj\projects\
samples\tune
```

2. You can optionally open the file Tune.java with a text editor to view some leJOS code. Compile the sample Java file:

```
nxjc Tune.java
```

3. This creates a file called Tune.class, which you can check by typing dir. Now it is time to upload your code to the NXT brick. Plug in your USB cable (we won't get into Bluetooth quite yet). Turn on the NXT by pressing the orange button and type:

```
nxj —r Tune
```

Just for fun!

Did you know you are a cross platform developer? Normally when developing an application you program on one platform (such as Windows) and run your code on the same platform. With leJOS NXJ, you are programming on one platform, such as Windows, and uploading the code to another platform, the NXT brick. So you can now boldly go out and state that you are a cross platform developer.

After a moment you should hear a tune play from your NXT brick.

CHAPTER 2

IDE

TOPICS IN THIS CHAPTER

- Setting up a Development Environment
- leJOS Eclipse Plugin

Java programming is possible with a text editor and a command line. However, it's easier to click on buttons to make things happen rather than typing commands and optional parameters. Standard text editors included with Windows and Macintosh systems do not have many features to help you enter code. They won't tell you when you've misspelled the name of a class or forgotten a bracket.

An IDE, or *Integrated Development Environment*, is a tool that allows you to enter, compile, and upload code to your NXT using simple buttons. It also monitors code syntax, coloring your code so you can more easily identify keywords and variables. This section will suggest a free, open source IDE for your leJOS NXJ needs.

One of the best open source IDEs is Eclipse by IBM (see Figure 2-1). It's free, powerful, and easy to use. It makes sense to use an advanced IDE like Eclipse with the NXT since your code can grow quite large.

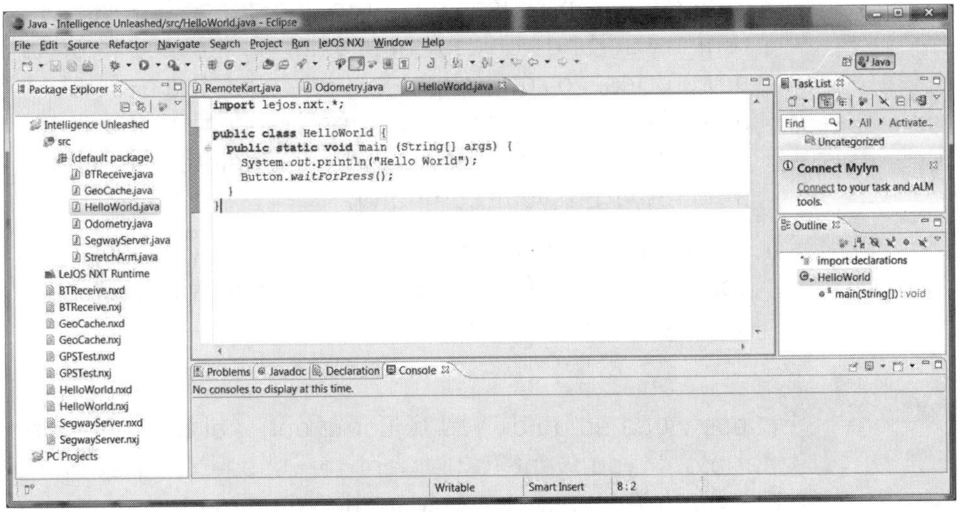

Figure 2-1: Programming in Eclipse

<u>Setting Up Eclipse</u>

The version used for this book is Eclipse Indigo, released in June, 2011. I suggest using the most recent version of Eclipse, unless you encounter problems with the leJOS plugin. The loading screen of Eclipse tells you what version you are using.

1. Download Eclipse from: www.eclipse.org in the download section. There are several different distributions of Eclipse for different types of users. The most basic one is the Eclipse IDE for Java Developers. Make sure to download the 32-bit version even if you are using a 64-bit computer, because our leJOS plug-in will not work with the 64-bit version.

2. The download is a zip file. Decompress the files into a directory. This will be the permanent location for Eclipse.

3. That's it. Eclipse uses no setup and doesn't store registry settings or copy native library classes to other directories. To run Eclipse, double click the executable file in the Eclipse directory (or create a shortcut to this). To uninstall Eclipse, merely delete the Eclipse directory from your computer.

4. The first time you run Eclipse it will ask for a workspace location. Your best option is to go with the default one it gives to you.

Eclipse will also guide you to some optional help files and tutorials. If you want to go right to Eclipse programming, click Go to Workbench or close the Welcome tab.

It's a good idea to have Eclipse automatically search for updates, in case there are software patches or new features. Click on Window > Preferences. Double click Install/Update in list, and highlight Automatic Updates (see Figure 2-2). Place a checkmark next to Automatically find new updates and notify me and click OK. This will update not only Eclipse but your plugins as well.

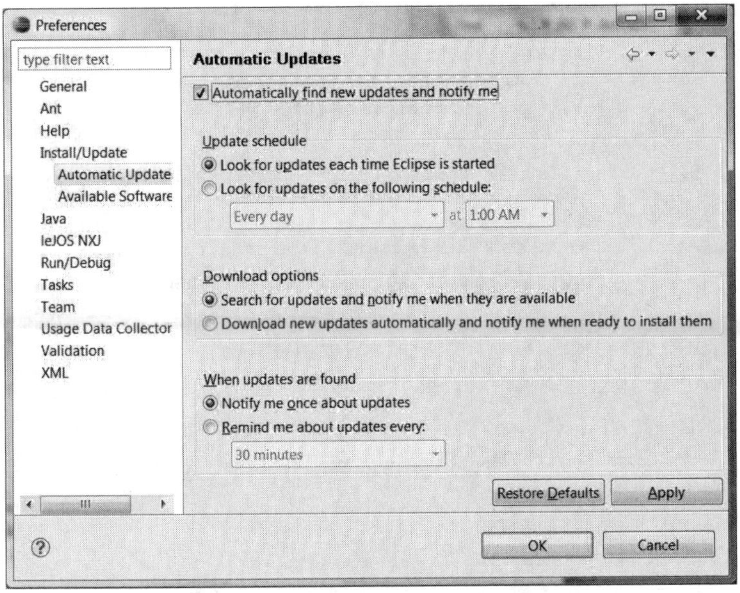

Figure 2-2: Automatically updating Eclipse

Using Eclipse with leJOS NXJ

Now that you have Eclipse installed it's time to install the leJOS Plugin.

1. In Eclipse, select Help > Install New Software... (see Figure 2-3).

Figure 2-3: Selecting new software to install

2. You will receive a dialog to input a URL. Click add and you will see another dialog box (see Figure 2-4). Enter the name "leJOS NXJ" and for the location enter the following: http://lejos.sourceforge.net/tools/eclipse/plugin/nxj/

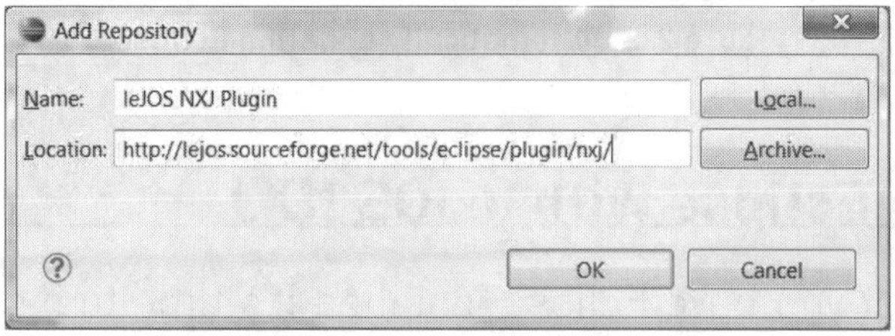

Figure 2-4: Adding a repository

3. Click OK. You should see a new item in the main box. Place a checkmark in the box next to the new item and click Next (see Figure 2-5).

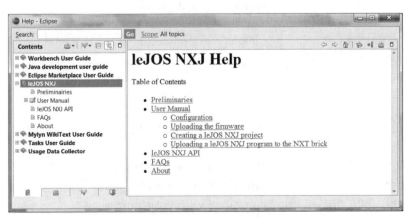

Figure 2-5: Entering the leJOS NXJ Plugin location

4. Read and accept the license agreement, and click Next. The plugin will automatically install.

5. When complete, it will ask to restart Eclipse. Once it has restarted, you will see some subtle changes with Eclipse. The plug-in will add new leJOS menu items to a variety of places within Eclipse. Let's explore the new options.

Help Files

In case you ever need a helping hand, there is a section in the help files specifically for leJOS NXJ. These files are continually updated as the plug-in is updated, so they are the final word on what you need to do to use the latest plug-in. Select Help > Help Contents and you will see a section for leJOS NXJ (see Figure 2-6). Feel free to browse them at any time.

Figure 2-6: Browsing the help files for leJOS NXJ

Configuring leJOS in Eclipse

Eclipse will automatically look for the NXJ_HOME environment variable to locate leJOS NXJ. Let's check to make sure the preferences are to your liking. Select Window > Preferences and then select leJOS NXJ from the list (see Figure 2-7). If the leJOS NXJ directory is incorrect, either type in the location of browse to it. Make sure you browse to the main directory and not one if its subdirectories.

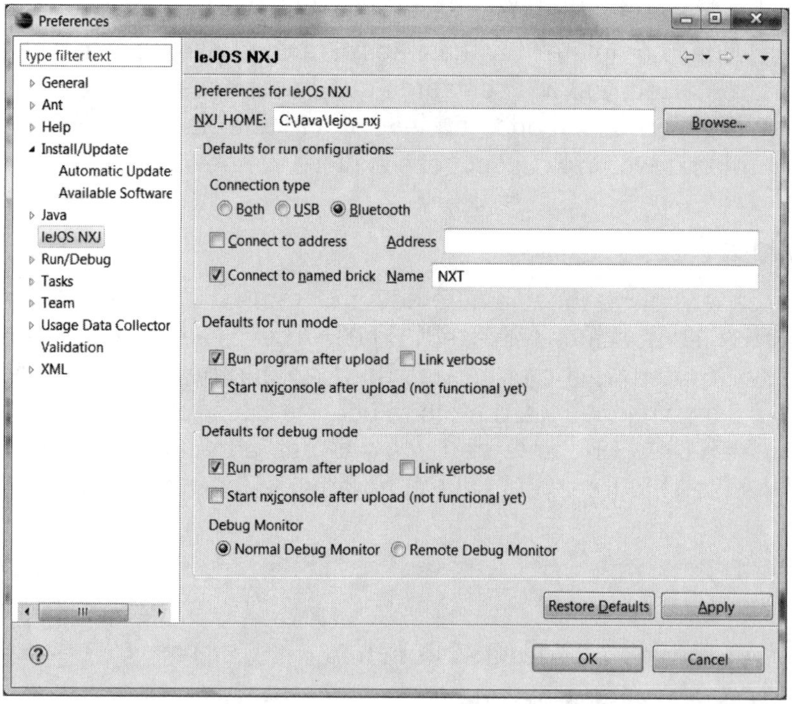

Figure 2-7 Selecting preferences

You can also select other options according to your own tastes. If you have a Bluetooth adapter, I recommend checking Bluetooth. You can also enter the name of your brick if there are other bricks in the vicinity. If you leave this field blank, leJOS will seek out any bricks paired with your computer.

Project Wizards

Now let's create a place to enter some code. Eclipse keeps individual Java projects in their own project directories. For example, if you are creating a large, multi-class project dealing with mapping, you would create your own project with its own directory to store the class and data files. Let's try creating a project which we will use to store the code in this book.

1. Select File > New > Project...

2. In the next window, double-click LeJOS to expand the folder options (see Figure 2-8). We want to create a leJOS NXJ Project (not a leJOS PC Project—we'll cover that topic later). Select leJOS NXJ Project and click Next.

3. For the Project Name, enter Intelligence Unleashed. Then click Finish.

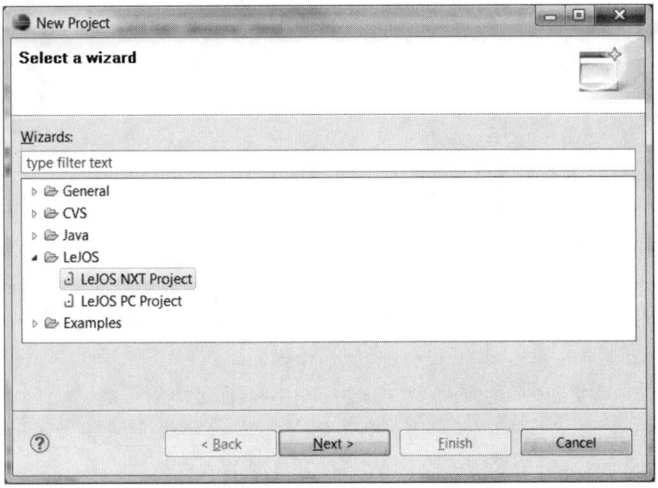

Figure 2-8: Using the project wizard

You can create as many projects as you want. You might want another project for your own programs, so go ahead and create another right now if you like.

Creating and uploading a program

Let's try entering a simple Hello World program using Eclipse and then uploading it using the leJOS NXJ plug-in.

1. To add a new class file, select File > New > Class (see Figure 2-9). Enter HelloWorld in the Name field. Eclipse also offers other options, such as automatically adding a main() method. Check this if you want Eclipse to do some of the typing for you.

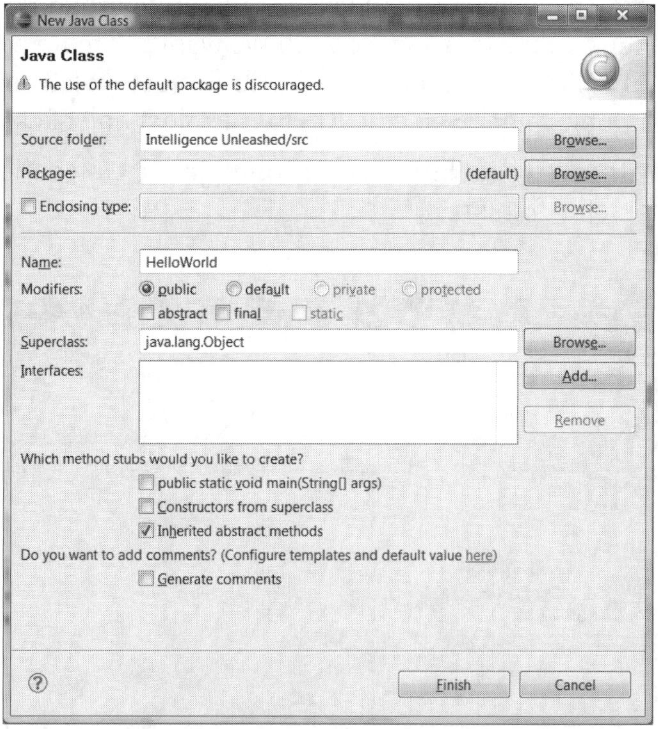

Figure 2-9 Creating a new Java class file

2. Click Finish when you are done and you should see a new class file with some starter code.

3. Enter the HelloWorld code below into the file.

4. Click the save button, turn on your NXT brick, then click the green Run button in the Eclipse toolbar. A

popup window will appear the first time you clikc this for a class file (see Figure 2-10). Select leJOS NXJ Program and click OK. The program will begin uploading.

Figure 2-10: Selecting how to run a class

5. Your NXT will beep when it is uploaded and automatically run, assuming you are using the default settings (see Figure 2-11). If you do not want the program to run automatically every time it is uploaded, select Window > Preferences > leJOS NXJ and uncheck Run Program after Upload.

```
import lejos.nxt.*;
public class HelloWorld {
  public static void main (String[] args) {
    System.out.println("Hello World");
    Button.waitForPress();
  }
}
```

Run Configuration

You can also configure the profile for each class you are working on. Instead of clicking the green button, click the arrow next to it and select Run Configurations... Select the configuration for Hello World (see Figure 2-12). This allows you to set some projects to run automatically and some to allow you to run them when you are ready.

Figure 2-11: Viewing the HelloWorld program

![Run Configurations dialog]

Run Configurations

Create, manage, and run configurations

type filter text

- Java Applet
- Java Application
- JUnit
- LeJOS NXT Prog
  - BTReceive
  - GeoCache
  - HelloWorld
  - SegwayServe
  - StretchArm
- Task Context Te:

Filter matched 13 of 1:

Name: HelloWorld

| Main | Classpath | Source | Common |

Project:

Intelligence Unleashed   Browse...

Main Class:

HelloWorld   Search...

When in run mode:

☑ Run program after upload   ☐ Link verbose

☐ Start nxjconsole after upload (not functional yet)

When in debug mode:

☑ Run program after upload   ☐ Link verbose

☐ Start nxjconsole after upload (not functional yet)

◉ Normal Debug Monitor   ◉ Remote Debug Monitor

Apply   Revert

Run   Close

Figure 2-12: Choosing run configurations

Custom Eclipse Tools

You can also set up your own custom tools with Eclipse. Anything that is executable from the command line is capable of being a tool, such as a notepad.

1. Select Run > External Tools > External Tools Configuration... to bring up a new window.

2. Select the Program item and click the New button (see Figure 2-13).

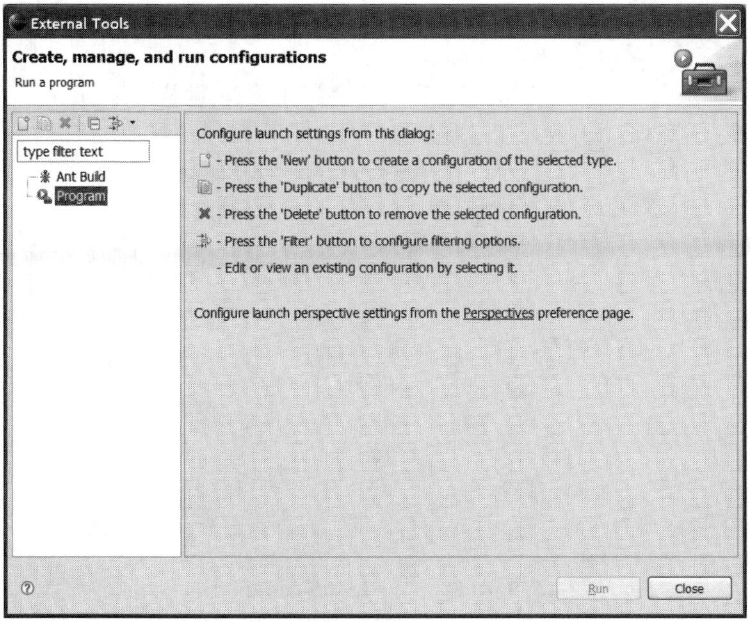

Figure 2-13: Adding external tools to Eclipse

3. For this example we will set up the leJOS NXJ image converter, located in the bin directory of your NXJ install. In the name field, enter Image Converter (see Figure 2-14). For location, click Browse File System and browse to \bin\nxjimage.bat.

4. Click on Run and the Image Converter tool should launch. From now on, you can launch this tool from Eclipse with the External Tools button (it looks like a small green play button with a briefcase).

No other settings are necessary for the Image Converter, but you could optionally include run-time parameters in the arguments field of the configuration (see Figure 2-14).

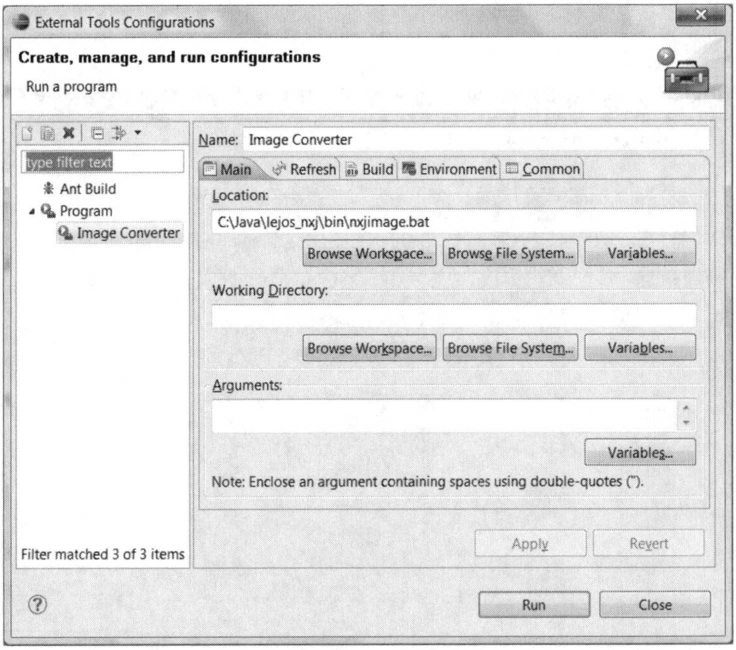

Figure 2-14: Setting up the leJOS compiler in Eclipse

Eclipse Summary

So what is so good about Eclipse? Try making an error in your code. Notice that right away it tells you where there is a problem? It lists all the errors or warnings, shows you exactly where they are, and even tells you what is wrong and how to remedy the problem. When you type an object name and press the period key it even pops up all the methods you can access.

Although Eclipse is simple, it has many tools to make your coding experience easier. Briefly, click on the Source menu item and look at some of the tools. Cleanup lets you import code and clean it up to your preferred style, or clean up your own code. If your indentations are wrong, highlight your code and select Correct Indentation. Instantly it fixes everything. There are also functions for generating try-catch blocks and getter/setter methods.

Sometimes you need to overhaul your code in major ways. Check out the Refactor menu item, which you can use when you want to rename a class. If you want to rename a variable, right click it and select Refactor > Rename. You can even overhaul the architecture of your code, such as making an interface from a class. These methods go through your entire code making changes.

> **TIP:** Eclipse can be enhanced using a variety of plugins. For example, it can be difficult and time consuming to create a graphical user interface (GUI) using raw Java code. You can download a plugin that allows you to graphically design a GUI for your program which then automatically generates the code. Most plugins are free for non-commercial use. Eclipse plugins can be found at: www.eclipseplugincentral.com
>
> Plugins are installed through Eclipse (Help > Software Updates > Find and Install...) however you need to enter the remote location of the plugin supplier. The remote location is typically located on the web page for the project.

CHAPTER 3

Motors

TOPICS IN THIS CHAPTER

- Motor specifications
- RJ12 Cables
- Carpet Rover
- Regulated NXT motors
- Unregulated NXT motors
- RCX motors
- Multiple Motors
- Inverting Motors

Motors are the source of all movement in the NXT kit. Without them, your LEGO robots would be motionless statues. The motors play a central role in all robotic projects, and for this reason, the leJOS developers have spent a lot of time and effort perfecting the motor classes.

If you ask the motor to rotate 180 degrees, the motor algorithm will precisely rotate the angle without overshooting the target. One of our developers even went as far as testing performance by hanging a heavy mass from a pulley wheel over the edge of the table so that even under load, the motors will come to a halt where you want them to.

You can also control the motor speed, even if the robot is travelling up or down inclines. When you specify a speed, the internal code will moderate the power level to maintain a constant speed. You can even query the Motor class at any time to see how fast the motor is rotating (more on this later).

To combat slippage when a vehicle begins to move, the motor classes even allow acceleration. This means the wheels of your robot can start turning slowly and gradually accelerate to full speed, minimizing skidding and resulting in more accurate movement.

All of this sophisticated functionality is owed to the optical encoders embedded in the NXT motors, which keep track of axle rotation. By monitoring the motor position over time and adjusting the power to each motor, controlled speed and acceleration are achieved. The motors can turn in a direction for thousands of rotations and come back to the exact starting position at any time. This feature opens up incredible possibilities for robot creation, especially with navigation and robot arms.

Internally, the NXT *servo motor* contains gears that reduce the speed of the motor and increase power (see Figure 3-1). This gives the motors an unusual shape.

Figure 3-1: Inside the LEGO NXT Motor (photo provided by LEGO Education)

Cables

The NXT cables contain six wires, making them somewhat rigid. The NXT kit contains seven cables in total (one for each port) with three different lengths (see Table 3-1). If any of these prove too short, there are other lengths available from Mindsensors.com.

Short	Medium	Long
20cm (8.5 inches)	35cm (14.5 inches)	50cm (18.5 inches)

Table 3-1: Comparing cable lengths

LEGO NXT uses a connector known as RJ12, which looks much like a phone connector (see Figure 3-2). Since the connectors for sensors and motors are identical, you might think you can hook up and motor to a sensor port. This does not pan out, unfortunately, because the wire signals are different.

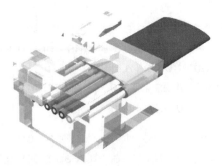

Figure 3-2: Viewing the RJ12 connector

Carpet Rover

Before examining the details of the motor classes, let's create a project to show off some motor functionality. Our first project is a simple robot to familiarize you with building and programming. The following section contains building instructions for a robust robot called Carpet Rover (see Figure 3-3). This robot will be used for many of the chapters dealing with navigation, so when you are done with this project, don't take it apart quite yet.

Figure 3-3: Carper Rover robot

TIP: The following LEGO plans are easier to understand if you first gather up the parts for the *current* step (shown in the upper left corner) and then assemble the step. If you pick parts one at a time out of your box it will take longer and you might lose track of whether you've used the correct number of parts.

1

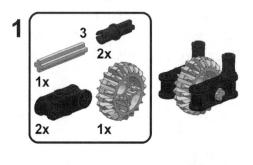

2

3

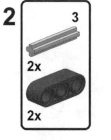

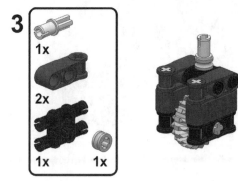

4

2x

1x

5

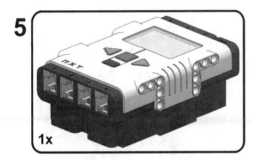

1x

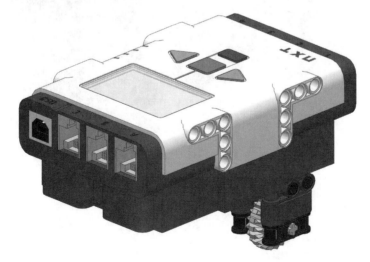

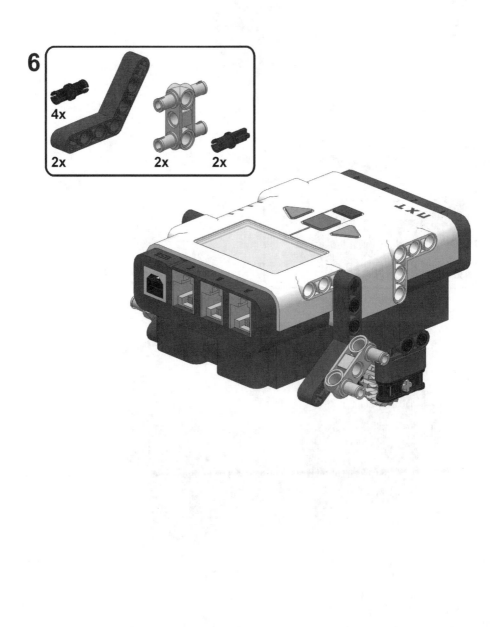

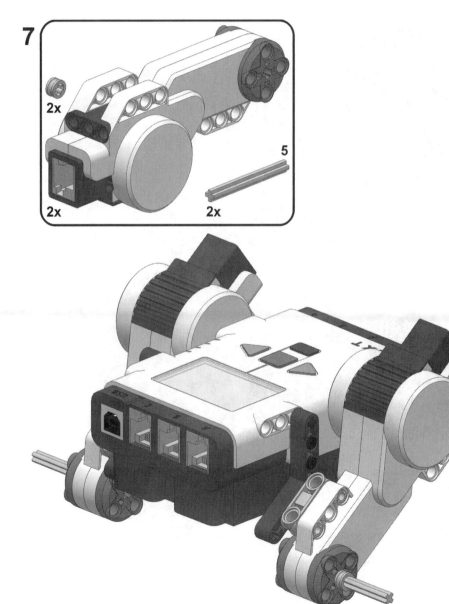

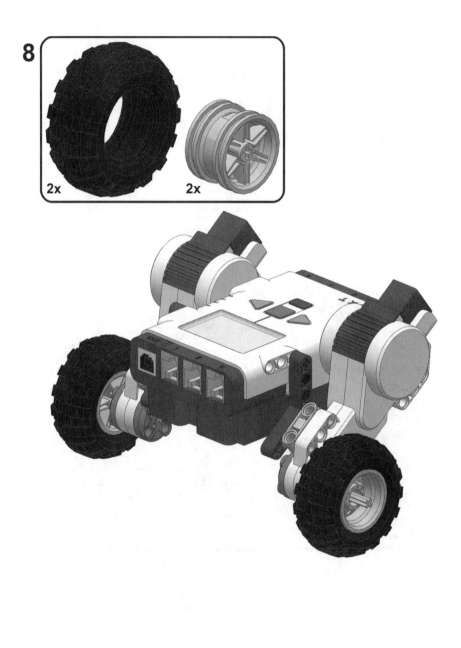

8

2x   2x

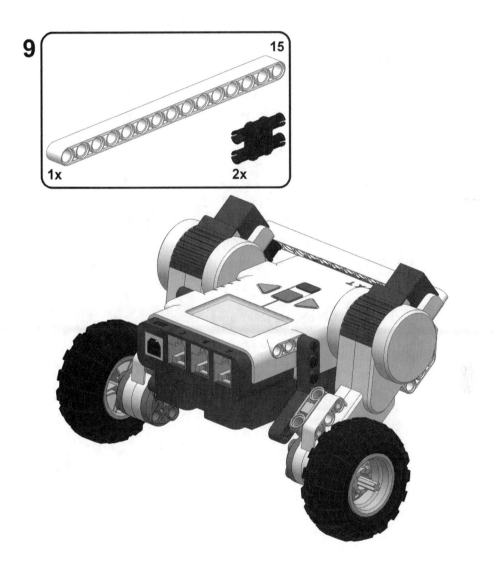

To finish the model, plug two medium cables into ports B and C, and then attach them to the motors. The robot is not very smart at the moment, so let's upload some code to make him move.

Twirler

This mesmerizing project will have your robot spinning around the room performing a graceful ballet. The acceleration value is set very low, so you will notice a gradual transition in motor speed between moves.

To enter the code in Eclipse, create a new class in your project by selecting File > New > Class (see Figure 3-4). Type the class name as Twirler and click Finish. If you are using command line, open a text editor and save the file to a directory. You will need to create a directory to run your code from.

Figure 3-4: Starting a new class in Eclipse

NOTE: I strongly recommend downloading all the code in this book from the book's website. This code is continually updated in case leJOS NXJ changes and it saves you from typing the entire code yourself. The website is at: www.variantpress.com

```java
import lejos.nxt.*;

public class Twirler implements
ButtonListener {

  public static void main(String [] args)
    throws Exception {
    Twirler listener = new Twirler();
    Button.ESCAPE addButtonListener(listener);
    Motor.B.setAcceleration(500);
    Motor.C.setAcceleration(500);
    Motor.B.forward();
    while(true) {
      float speed1 = (float) (Math.random() *
800);
      float speed2 = (float) (Math.random() *
800);
      Motor.B.setSpeed(speed1);
      Motor.C.setSpeed(speed2);
      if(Math.random() < 0.5) {
        Motor.C.forward();
      } else {
        Motor.C.backward();
      }
      Thread.sleep(5000);
    }
  }
  public void buttonPressed(Button b) {}

  public void buttonReleased(Button b) {
    System.exit(0);
  }
}
```

Compiling and Uploading in Eclipse

1. Enter the Twirler code into the class file you opened above and save it.

2. You might want to turn off the automatic run default in Eclipse, especially if you are using a USB cable rather than Bluetooth. To do this, select Windows > Preferences. Click leJOS NXJ and uncheck "Run program after upload". This will prevent your robot from taking off with the USB cable still attached.

3. Now turn on your NXT brick and plug in the USB cable (if appropriate).

4. Press the green run button. The first time, you will need to select "leJOS NXT Program".

Compiling and Uploading from Command Line

If you want to compile from the command line without Eclipse, use the commands below.

1. From a command line prompt, change to the directory in which you saved the above code.

```
cd \java\nxj_code
```

2. Compile the Java file:

```
nxjc Twirler.java
```

3. Now upload the file to the NXT. We'll use Bluetooth for this example, but if you only have the USB cable, substitute the "b" with a "u" below. Turn on the NXT by pressing the orange button and type:

```
nxj -b -n Twirler
```

Running the Program

Use the grey arrow keys on the NXT brick to navigate to the Files menu option (see Figure 3-5). Press the orange enter button, then select the Twirler filename. Set the robot down on the floor and then select execute. When you are done, hit the grey escape button to end the program.

Figure 3-5: Selecting the file menu

Hopefully you noticed the effect of setting a low acceleration. This robot has no ability to sense objects, so it will eventually crash given enough time. We'll try to fix it up with a sensor in the next chapter, but for now let's examine the motor classes in more detail.

> **TIP:** You can always stop code on the NXT by pressing the orange enter button and the grey escape button at the same time. This is much like using Ctrl-Alt-Delete to end programs in Windows.

Regulated Motors

The Motor class is your main entry point for accessing regulated NXT motors. This class contains three static instances: Motor.A, Motor.B and Motor.C. Regulated simply means the code in this class controls speed, acceleration and stopping. Let's examine the major methods.

lejos.nxt.NXTRegulatedMotor

```
public final boolean isMoving()
```

Returns true if the motor is in motion.

```
public final boolean isStalled()
```

Indicates if the motor is obstructed or under too much load and unable to rotate.

```
public final void stop()
```

Causes the motor to stop instantly. Once it is stopped, it resists any further motion.

```
public void flt()
```

Causes the motor to lose power and glide to a stop.

```
public void rotate(int angle)
```

Causes the motor to rotate the desired angle (in degrees).

```
public void rotateTo(int limitAngle)
```

Causes the motor to rotate to limitAngle. The tachometer should be within 1-2 degrees of the limit angle when the method returns.

```
public void setAcceleration(int acceleration)
```

Enables smoother acceleration. Motor speed increases at the desired rate, in degrees/second^2. 6000 is the default.

```
public final void setSpeed(int speed)
```

Sets motor speed, in degrees per second. Up to 900 degrees per second are possible with fully charged batteries. The actual maximum speed of the motor depends on battery voltage and load.

```
public final int getSpeed()
```

Returns the set motor speed, in degrees per second. Does not return the actual measured speed—see getRotationSpeed().

```
public int getLimitAngle()
```

Returns the angle (in degrees) that a Motor is rotating to.

```
public int getRotationSpeed()
```

Returns actual speed, in degrees per second. A negative value means the motor is rotating backward.

```
public int getTachoCount()
```

Returns the tachometer count in degrees.

```
public void resetTachoCount()
```

Resets the tachometer count to zero.

```
public void forward()
```

Causes the motor to rotate forward.

```
public void backward()
```

Causes the motor to rotate backwards.

> **NOTE:** There are overloaded methods for many of the methods shown above. These methods have an optional boolean parameter called "immediateReturn". Normally the method does not return until it is done performing the rotation, but if you set this value to true, the method will return immediately even though the motor has not yet reached the target. This is often useful if you want two motors to simultaneously rotate to a target angle, as follows:
>
> ```
> Motor.A.rotateTo(500, true);
> Motor.B.rotateTo(-500);
> ```
>
> **TIP:** NXT robots have a hidden sensor to detect when they've hit an obstacle: the motors. That's because the Motor class has an isStalled() method. If the motor is turning slower than it is supposed to, it means there is probably an obstacle hindering the robot and slowing down the motors.

MotorPort

The MotorPort class was added to the leJOS NXJ to handle backward compatibility with RCX motors, and to allow for multiplexer units which extend the number of motor ports. There are three static instances of MotorPort: A, B, and C. You can use one of these to construct motors, such as RCX-Motor (see below).

```
RCXMotor m = new RCXMotor(MotorPort.A);
```

You are unlikely to call any of the methods in MotorPort directly with regular robotics projects, but if you are curious you can browse the leJOS NXJ API documentation.

Unregulated Motors

Sometimes a project requires direct access to motor power rather than using the speed regulation code. For example, the Segway-like robot featured in this book bypasses the regulated code in order to access power directly. The class implements the BasicMotor interface, making it almost identical to the RCXMotor. The only exception is that NXT-Motor has access to two tachometer methods:

- getTachoCount() – returns the tachometer reading
- resetTachoCount() – resets the tachometer to zero

NXTMotor

```
public void setPower(int power)
```

Sets power to the RCX motor.

Parameters: power - power setting (0 – 100)

```
public int getPower()
```

Returns the current power setting (0 – 100).

```
public void forward()
```

Causes motor to rotate forward.

```
public void backward()
```

Causes motor to rotate backwards.

```
public boolean isMoving()
```

Returns true if the motor is in motion.

```
public void flt()
```

Causes motor to float. The motor will lose all power, but this is not the same as stopping.

```
public void stop()
```

Causes motor to stop instantaneously. The motor will re-sist any further motion.

RCXMotor

RCXMotor is a class for legacy RCX motors. The methods in RCXMotor are identical to those in the unregulated NXT-Motor class because they both implement the BasicMotor interface, so we will not bother reviewing the methods again. To obtain an instance of this class, use a static instance of MotorPort in the RCXMotor constructor.

lejos.nxt.RCXMotor

```
public RCXMotor(BasicMotorPort port)
```

Constructor for RCXMotor.

RCXMotorMultiplexer

The RCXMotorMultiplexer class allows access to the Mind-sensors.com multiplexer for RCX motors. It is similar to RCXLink in that it contains instances of RCXMotor (four in this case). You can access these motors as follows:

```
RCXMotorMultiplexer expansion = new
RCXMotorMultiplexer(SensorPort.S1);
expansion.A.forward();
```

The methods to control the each RCX motor are the same as those found in RCXMotor.

Inverting Motors

The direction of rotation is sometimes confusing with NXT motors. Look at the motor in Figure 3-6. By default, when you ask the motor to rotate forward, it will rotate counter-clockwise for the position shown in Figure 3-6. However, sometimes you build a robot and when you ask the motors to rotate forward they end up rotating in the opposite direction you wanted.

To make the motors go in the correct direction, you can make a new motor that is a mirror image of the original motor using the MirrorMotor class in the lejos.robotics package. A factory method called invertMotor() allows you to retrieve an inverted motor. For example, to reverse the direction of Motor.B, use the following line of code.

```
RegulatedMotor b =
MirrorMotor.invertMotor(Motor.B);
```

The instance b will now rotate in the opposite direction, and all tachometer readings are also reversed. You can also get a reverse direction of an unregulated motor using the over-loaded invertMotor() method as follows.

```
EncoderMotor c = new NXTMotor(MotorPort.C);
c = MirrorMotor.invertMotor(c);
```

You will get a chance to use inverted motors later in this book.

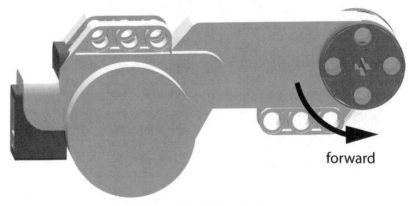

forward

Figure 3-6: Motor rotation

If the motors are the most important component for performing moves, the sensors are the most important components for performing *informed* moves. Picture a robot that travels around your house. In the past, such an activity was short-lived because the robot invariably became stuck. It would tip over, run into a wall without the bumper activating, or the wheels would become stuck on some low-lying object.

Now picture a robot that wanders around your house avoiding objects with the distance sensor. If the sensor fails to detect an object (which it often does when it approaches a wall at an angle), the robot can still tell if the wheels are stuck by monitoring decreases in rotation speed. If the robot tips over it uses a tilt sensor to identify the problem. It can then try to right itself. Such a robot could be left alone for hours, and when you return it would still be exploring your house. This is all possible by using the sensors available to LEGO NXT robots.

In this chapter, we'll take the Twirler robot from the previous chapter and add an ultrasonic sensor to make it more aware of its surroundings as it moves around the room. But first, let's examine the sensors and their classes.

Sensors

The first thing people notice about LEGO sensors is that they are identical in size and shape, except for the ultrasonic sensor which looks like a robot head (it is wider at the front, but the rear is identical to the other sensors). All of the sensors use the same means of attaching to the robot (three holes for friction pins or axles), making them highly interchangeable.

All of the sensors plug into one of four sensor ports. In your code, the sensor classes must indicate which port they are plugged into. They do this using the SensorPort class.

lejos.nxt.SensorPort

The SensorPort class is similar to MotorPort. There are four static SensorPort objects in this class. Generally these are used in the constructors for sensors, as follows:

TouchSensor ts = new TouchSensor(SensorPort.S1);

SensorPort contains many methods for accessing data directly from the port, either using the I^2C protocol or analog sensors. You are unlikely to use these methods yourself since you are merely passing the SensorPort to a sensor class, but if you are interested in developing your own sensor classes you might want to examine the SensorPort API more thoroughly.

Generic Sensor Interfaces

The LEGO NXT has a mini-economy of sensors manufactured by third parties. For example, compass sensors are produced by both HiTechnic (marketed through LEGO) and Mindsensors. Although the individual compass sensors are designed and programmed differently, they both return direction readings.

On the leJOS NXJ project, we wanted our API to be compatible with sensors, even if they came from different manufacturers. To accomplish this, we created a series of interfaces that describe the main functions of each family of sensors. For example, we created an interface called DirectionFinder which can return a standard direction value. Then we created classes for the HiTechnic sensor and Mindsensors sensor, both of which implement the DirectionFinder interface.

So what does this accomplish? Now a class can accept either of these sensors in a method or constructor. The following code sample shows a sample class with a constructor that accepts a DirectionFinder.

```
import lejos.robotics.DirectionFinder;

public class MyRobot {
  private DirectionFinder df;

  public MyRobot(DirectionFinder compass) {
    this.df = compass;
  }
}
```

None of the sample code is functional yet, but as you can see, the class will now accept any direction finder that exists or comes along in the future. This effectively future-proofs our API to allow people to add new sensors without us having to reprogram the other classes in leJOS NXJ.

The lejos.robotics package also includes other sensor types, such as RangeFinder, which is implemented by the UltrasonicSensor class. You can make your own sensors that implement these interfaces. For example, my previous book Core LEGO Mindstorms Programming contained projects to build a compass sensor and a range sensor. If you have an RCX to NXT adapter then you can program your own class to implement the DirectionFinder and Range-Finder interfaces, and thus allow your homebrew sensor to work with the leJOS NXJ API.

In fact, the DirectionFinder does not have to be a compass sensor. Other sensors can determine direction in other ways, such as the GyroDirectionFinder located in the lejos. nxt.addon package.

You can even experiment with exotic methods for determining direction. Imagine a sensor that estimates direction based on the position of the sun, given that it knows the latitude, longitude, and current time. Or more simply, it could estimate direction based on spotting a fixed light source in a room by using a light sensor mounted on a motor.

The different sensor types are located in the lejos.robotics package. You can probably guess what they do based on their names:

- Accelerometer
- ColorDetector
- DirectionFinder
- Gyroscope
- LightDetector
- PressureDetector
- RangeFinder
- Tachometer
- Touch

Analog Sensors

Analog sensors are sensors that return values based on the voltage returned by the sensor. These values typically range from 0 to 1023. Most of the sensors included in the LEGO kit are analog, including the touch sensor, sound sensor (NXT 1.0), color sensor (NXT 2.0), and light sensor (NXT 1.0).

Analog sensors return values very quickly, because there is no data bus to slow down the transmission of data from the sensor to the NXT. It simply needs to check the voltage value and return that value. Let's examine the analog sensors available in the NXT kit.

Just for fun!

According to ohloh.net, it has taken 34 person-years to develop leJOS NXJ and would cost almost two million dollars to program from scratch (with salaries of $55,000 per year).

Light Detectors

Both the NXT 1.0 and 2.0 kits contain sensors for detecting light, although the 1.0 sensor is not as advanced. The NXT 1.0 kit contains a single light sensor, which measures the intensity of light entering a tiny lens on the front of the sensor (see Figure 4-1). The sensor is also equipped with a red light-emitting diode (LED) which illuminates the scene in front of the sensor. The sensor can also detect light invisible to the human eye, such as infrared (IR) light emitted from a television remote control.

Figure 4-1: Light sensor

The light sensor is used to perform a variety of functions. By pointing the light sensor down, the robot can follow a black line. Sometimes the sensor is used to prevent a robot from driving off the edge of a table. This is possible because light values decrease significantly when an object, such as the floor in this example, is farther away (far objects do not reflect as much light as near objects). The light sensor can also distinguish dark objects from light objects, since dark objects reflect less light.

Light sensors have two modes:

- Active mode – the light sensor LED is illuminated. This is often used for line following or object detection.
- Passive mode – the light sensor LED is extinguished. This mode is used for ambient light detection, such as measuring the sun's brightness.

lejos.robotics.LightDetector

There are three classes that implement the LightDetector interface:

- lejos.nxt.LightSensor – the NXT 1.0 sensor
- lejos.nxt.ColorSensor – the NXT 2.0 sensor
- lejos.nxt.addon.RCXLightSensor – the legacy sensor

All of them use the basic LightDetector methods, although some of them have their own specialized methods that are unique. To use the LightSensor class, you must first create an instance of this class, as shown below:

```
LightDetector ld = new LightSensor(SensorPort.S1);
```

NOTE: All of the sensor constructors use SensorPort. However, if you look at the API it shows ADSensorPort or I2CPort. This architecture was adopted to differentiate between the two types of ports. SensorPort implements both of these types, which is why it is used in this chapter.

By default, the sensor is set to floodlit mode, which turns on the red LED. You can use an alternate constructor with a false Boolean value to keep the light off.

Once you have an instance of LightDetector, use the getLightValue() method to return a number between 0 and 100:

```
int brightness = getLightValue();
```

The brighter the light, the higher the light value. Without performing calibration, the values might not approach zero or 100. In order to spread out the values, you can calibrate the sensor.

Use calibrateLow() to set the zero level, and calibrateHigh() to set the 100 level. While executing calibrateLow(), make sure the sensor is completely in the dark. While executing calibrateHigh(), point the sensor lens at the brightest light in the room or at the sun on a clear day.

If you want more fidelity in your values, you can use the reeadNormalizedLightValue() method. This method allows more accuracy than readValue(), since it returns a larger scale. Values can theoretically range from 0 to 1023, but typically they range from 145 (dark) to 890 (sunlight).

Color Detector

The NXT 2.0 kit contains a single color sensor (see Figure 4-2). This sensor is similar to the light sensor, except that it can detect different colors. On top of that, the built-in lamp can emit red, blue, or green light from multicolored LEDs.

Figure 4-2: The NXT 2.0 Color Sensor

Another color sensor is available from HiTechnic. Both the official LEGO sensor and the HiTechnic sensor implement the ColorDetector interface. They can read red, green, blue (RGB) color values, or identify colors from a palette.

lejos.robotics.ColorDetector

To identify simple pre-defined colors, use the getColorID() method, which returns an integer representing a color constant.

```
int getColorID()
```

These values are accessed from the lejos.robotics.Color class. For example, Color.BLUE represents color values in the blue spectrum.

You can also retrieve RGB values as follows:

```
ColorDetector cd = new ColorSensor(SensorPort.S1);
Color color = cd.getColor();
int red = color.getRed();
```

The Color class can even identify the color spectrum via the Color.getColor() method, which returns a color constant from a palette of colors.

Touch Sensor

The touch sensor is the most basic sensor in the NXT kit (see Figure 4-3). It has a simple switch activated by the orange button on the front. The button has an axle hole, allowing a LEGO axle to attach directly to the switch. There is a single touch sensor in the NXT 1.0 kit and two touch sensors in the 2.0 kit.

Figure 4-3: Touch Sensor

The touch sensor returns a simple Boolean value indicating if the sensor is pressed or not. The TouchSensor class implements the Touch interface, which contains one simple method.

```
boolean isPressed()
```

As the name indicates, this method checks if the sensor is pressed.

Sound Sensor

The NXT 1.0 kit contains a single sound sensor (see Figure 4-4). Although it resembles a microphone, it really just measures the loudness of sound in decibels (dB). You can't use the sound sensor to record sound files to the NXT brick. Since sound is louder when the source is near, the sound sensor allows robots to home in on sound sources. It can also react to sounds, such as clapping.

Figure 4-4: Sound sensor

The SoundSensor class has two modes of operation, dB and dBA mode. The dB mode measures straight decibels, while dBA measures sound intensity weighted for the range of human hearing. These are explained below.

lejos.nxt.SoundSensor

Before you can use a sound sensor, you will need to create an instance of the class as follows:

```
SoundSensor s = new SoundSensor(SensorPort.S1);
```

By default, the sensor will be set to DB mode. You can use the alternate constructor to set it to DBA mode by default. You can also set the mode to dBA by using setDBA():

```
s.setDBA(true);
```

To read the current decibels, use readValue():

```
public int readValue()
```

This method returns the current sensor value as a percentage (0 to 100).

I²C Sensors

All four NXT sensor ports allow a standard protocol called *Inter-Integrated Circuit* or I²C (pronounced I-squared-C). I²C is a bus that allows transmission of data to and from sensors. Philips invented the standard in the early 1980s and since then it has seen some use in cell phones and other small devices.

In contrast to the analog sensors which dominate in the LEGO NXT kit, most of the third-party sensors available for the NXT are I²C.

Although there are four physical ports, the I²C ports are capable of using far more than four sensors at once. As long as the sensor uses Auto Detecting Parallel Architecture (ADPA), you can connect additional sensors using an expander.

I²C returns information slower than analog sensors, but it can return a richer set of data as opposed to a single value for analog sensors. Any reservations you have about I²C being too slow for robotics are unfounded, even for time-critical applications. High-speed I²C has been implemented for leJOS NXJ, which means return-times for data are around two milliseconds. The Segway-like robot in this book requires readings every eight milliseconds, so speeds are well within the requirements of even the most time-critical application.

Keep in mind, even though the I²C bus can return data very quickly, not all sensors can return data this fast. The ultrasonic sensor, for example, uses sound waves to ping objects, so its speed is limited by the time it takes to send and receive the ping. In other words, the ultrasonic sensor is more limited by the speed of sound than by the speed of the I²C bus.

lejos.nxt.I2CSensor

I2CSensor is the abstract superclass for dozens of I²C sensors, including UltrasonicSensor, CompassHTSensor, and ColorHTSensor. This means all the methods below can be called from these sensors. All sensors accept a SensorPort object in the constructor (e.g. SensorPort.S1). Unless you are making your own I²C sensors, the most useful methods are for reading sensor type, product ID and version numbers.

```
public int getData(int register, byte[] buf,
int len)
```

If you are making your own I²C class, use this to read data from the sensor.

Parameters:
```
register - I2C register, e.g 0x41
buf - Buffer to return data
len - Length of the return data

public int sendData(int register, byte[] buf,
int len)
```

Use this to send or change data on the sensor.

Parameters:
```
register - I2C register, e.g 0x42
buf - Buffer containing data to send
len - Length of data to send

public String getVersion()
```

Returns the sensor version number.

```
public String getProductID()
```

Returns the sensor product identifier. Usually this is the maker of the sensor, such as "LEGO".

```
public String getSensorType()
```

Returns the sensor type. For example, "Sonar" for the ultrasonic sensor

Ultrasonic Sensor

Even though the ultrasonic sensor looks like a pair of eyes, it actually has more in common with the sound sensor than a camera. The ultrasonic sensor sends out a sound signal (like a bat) that is nearly inaudible to humans, then measures how long it takes for the reflection to return. Since it knows the speed of sound, it can calculate the distance the signal traveled.

The ultrasonic sensor is the only I²C sensor included in the NXT kit. It measures distances to solid objects in centimeters or inches. The sensor is capable of measuring distances up to 255 centimeters, though returns are inconsistent at these distances because the return ping becomes weaker. The sensor is accurate from 6 to 180 centimeters, with objects beyond 180 centimeters not reliably located. It has an accuracy of plus or minus three centimeters, though the accuracy is better for close objects.

The ultrasonic sensor produces a sonar cone, which means it detects objects in front of it within a cone shape. The cone opens at an angle of about 30 degrees (see Figure 4-5). This means that at a distance of 180 centimeters the cone is about 90 centimeters in diameter. The cone shape is ideal for most robots, since it is better to scan a large area in front of the robot for possible collisions.

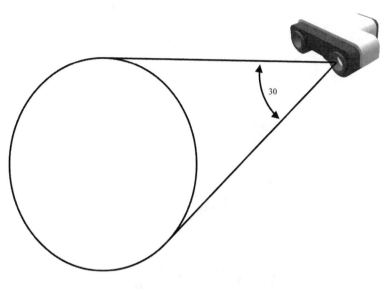

Figure 4-5: The 30 degree ultrasonic sonar cone

The ultrasonic sensor is a range sensor, so it implements the RangeFinder interface. The only other sensor to implement this interface is OpticalDistanceSensor, the class which accesses the high-precision sensors from Mindsensors. All of these sensors use the following method to obtain distances using the RangeFinder interface.

```
double dist = rf.getRange();
```

This method returns the distance to an object in centimeters. Some range finders, such as the ultrasonic sensor, are capable of returning multiple detected objects from one single scan. In that case, you can use the following method:

```
float [] dists = rf.getRanges();
```

Just for fun!

While the sensor is in use, place it next to your ear. You can hear rapid clicks continuously from the sensor, which sound almost like bat pings.

Now that we have some idea of how to use sensors, let's try extending the Twirler robot we created in the previous chapter. For this project, you will need to affix an ultrasonic sensor to the Carper Rover robot from the previous chapter, as shown in Figure 4-6.

Figure 4-6: Adding an ultrasonic sensor to the robot

1

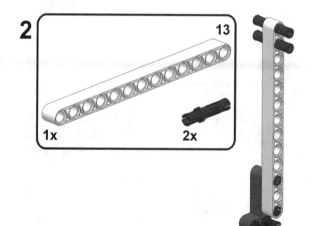

2 13

1x 2x

3

2x

1x

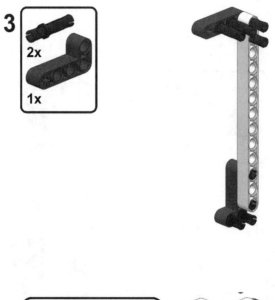

4

1x

1x

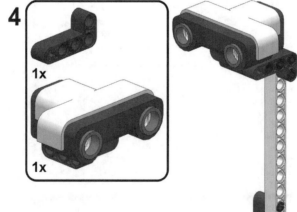

Plug the ultrasonic sensor into port 1. The code for this robot simply tells it to back up a certain distance if the range finder detects an obstacle. There are only about six new lines of code to add to the previous program. The entire program is shown below.

```java
import lejos.nxt.*;

public class TwirlerAware implements
ButtonListener {

public static void main(String [] args)
throws Exception {

  TwirlerAware listener = new TwirlerAware();

    Button.ESCAPE.addButtonListener(listener);

    UltrasonicSensor us = new
UltrasonicSensor(SensorPort.S1);

    Motor.B.setAcceleration(500);
    Motor.C.setAcceleration(500);
    Motor.B.forward();
    while(true) {
      if(us.getRange() < 40) {
        Sound.beep();
        Motor.B.setSpeed(800);
        Motor.C.setSpeed(800);
        Motor.B.backward();
        Motor.C.backward();
        Thread.sleep(4000);
        Motor.B.forward();
      }
      float speed1 = (float) (Math.random() * 800);
      float speed2 = (float) (Math.random() * 800);
      Motor.B.setSpeed(speed1);
      Motor.C.setSpeed(speed2);
      if(Math.random() < 0.5) {
        Motor.C.forward();
      } else {
        Motor.C.backward();
      }
      Thread.sleep(5000);
    }
  }

  public void buttonPressed(Button b) {

  public void buttonReleased(Button b) {
    System.exit(0);
  }
}
```

Keep in mind, the robot is only checking for obstacles every five seconds, and twirling around the whole time, so it is occasionally going to collide with objects. We'll find out later in the book how to make it perform better.

Direction Finders

Now that we've covered all the sensors by LEGO, let's examine some third party sensors. A compass can be useful for identifying direction. As previously discussed, both HiTechnic and Mindsensors have released compass sensors, which use the methods in the DirectionFinder interface.

lejos.robotics.DirectionFinder

To retrieve standard degrees, use the following method.

```
public float getDegreesCartesian()
```

This method retrieves compass direction. Compass readings increase clockwise, but Cartesian coordinate systems increase counter-clockwise. This method returns the Cartesian compass reading.

```
public void resetCartesianZero()
```

Changes the current direction the compass is facing into the zero angle. Affects only getCartesianDegrees()

```
public void startCalibration()
```

This method starts calibration for the compasses. When this is called, you must rotate the robot *very* slowly, taking at least 20 seconds per rotation. The Mindsensors compass requires at least two full rotations, while the HiTechnic requires one and a half to two rotations.

```
public void stopCalibration()
```

Ends calibration sequence.

The other class using DirectionFinder is GyroDirectionFinder. It requires a gyroscopic sensor, such as the HiTechnic gyro.

Accelerometer

There are two accelerometers available, one from Mindsensors.com and the other by HiTechnic. These sensors measure acceleration and tilt.

The following methods are available to the accelerometer sensors.

```
public int getXTilt()
```

Returns X tilt value.

```
public int getYTilt()
```

Returns Y tilt value.

```
public int getZTilt()
```

Returns Z tilt value.

```
public int getXAccel()
```

Returns positive or negative acceleration along the X axis as mg (mass x gravity). The variable g represents acceleration due to gravity (9.81 m/s). If this accelerometer was falling on earth and oriented along the X axis, it would return 9810 mg.

```
public int getYAccel()
```

Returns acceleration along the Y axis in mg.

```
public int getZAccel()
```

Returns acceleration along the Z axis in mg.

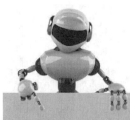

Just for fun!

If you have access to a sound sensor from the LEGO 1.0 kit, try making your own range sensor using the sound sensor and the NXT speaker. The theory is that the NXT speaker emits a short but loud ping. Once that is done, the sound sensor listens for the sound to echo off objects. You would, of course, need absolute silence for this to work—including no motor sounds. The speed of sound is 343.2 meters per second (at room temperature and dry air). Using this speed, we can estimate the distance to an object. You can use the RangeFinder interface for your new sensor.

CHAPTER 5

The JVM

TOPICS IN THIS CHAPTER

- The NXT Brick
- The Menu
- Batteries
- Buttons
- Speaker
- Javadocs

We've taken a look at motor and sensor support in le-JOS NXJ, but the NXT brick is capable of doing a lot of other things. This chapter explains how to use the other functions of the NXT brick. We will examine the Java Virtual Machine (JVM), its support of Java, and parts of the Application Programming Interface (API). But first, it would be a good idea to familiarize yourself with the NXT brick itself so you can understand some of the limitations of the JVM.

What's in There?

The NXT is 7.2 by 11.2 cm. With six AA batteries, NXT weighs 286 grams (160 grams without batteries). The durable plastic brick contains an Atmel® 32-bit ARM processor running at 48 MHz. This processor has direct access to 64 KB of RAM, and 256 KB of flash memory. The flash RAM stores programs and data even when there is no power, which saves battery life.

In an age of gigabytes, you might think 320 KB is not impressive. However, there are a few things that make this memory limitation irrelevant. First, robots don't use heavy-duty graphics or sound, which consumes heaps of memory in modern computers. Generally, only mapping projects and advanced artificial intelligence (AI) programs use significant memory.

For example, examine the specs for the Mars rovers *Spirit* and *Opportunity* (see Figure 5-1). Launched in 2003, each contained 256 KB of flash memory. If that's enough memory for NASA, you can trust it is enough memory for your own projects. If you do need a lot of memory for a program, you can control the NXT wirelessly from code running on your personal computer.

Figure 5-1: A Mars rover exploring terrain

Just for fun!

The brick also contains an Atmel AVR 8-bit processor running at 8 MHz to operate the servo motors and rotation sensors. This processor has access to an additional 4 KB of FLASH memory and 512 bytes of RAM. Why a separate processor just for the motors? Because the NXT needs to monitor the optical tachometer constantly in order to remain accurate. If this task was left to the main processor, it might fail to keep track of rotations while simultaneously doing something else, such as running your program.

NXJ Menu

The 1.0 version of leJOS NXJ contains a scrolling graphical menu system. It is pretty self explanatory. Basically you navigate using the left and right arrows, and make selections with the orange enter button. To back up from a menu section, press the dark grey escape key. Or, if you are at the main menu, you can turn off the NXT with the same button.

Figure 5-2: The graphical menu system

Batteries

The NXT brick uses six 1.5 volt AA batteries which provide 9 volts. However, if you frequently use your NXT, disposable AA batteries will cost you a small fortune. There are two off-the-shelf options for rechargeable batteries.

The first option is the rechargeable lithium ion battery from LEGO (see Figure 5-3). This battery provides at least 7.4 volts (closer to 8.2 volts after recharging). It fits into the regular battery case, but it also increases the depth of the NXT brick slightly.

LEGO also sells an AC adapter for charging the lithium ion battery which, conveniently, can be done while it is still inside the NXT brick. Conceivably, this means you could devise a robot that drives up to a recharging station when the batteries are low, although this would require some hardware hacking.

Lithium batteries provide power even while they are being charged, meaning your robot can also feed directly from household current. People who want to create robots that operate 24 hours a day, seven days a week (such as an Internet controlled robot) will find the lithium battery a necessary accessory.

Figure 5-3: LEGO Lithium Ion battery

Another option is to use six rechargeable AA batteries. There are two common types: Ni-MH (Nickel Metal Hydride) and Ni-Cd (Nickel Cadmium). Both work well, but the Ni-MH battery supplies 1.2 volts while the Ni-Cd battery supplies 1.25 volts. They don't store as much charge between recharging as lithium, however (see Table 5-1).

> **WARNING:** Rechargeable batteries provide 7.2 to 7.5 volts to the NXT, which means your motors will not operate as fast or powerfully as they would with 9 volts. Other than that, rechargeable batteries work well.

	Rechargeable?		Duration
Alkaline	No	9.0	longest
Ni-Cd	Yes	7.5	lowest
Ni-MH	Yes	7.2	40% more than Ni-Cd
Lithium	Yes	7.4	2 x Ni-Cd

Table 5-1: Comparing battery options

The leJOS menu monitors battery charge. It provides a simple battery icon on the main menu which indicates remaining charge in the battery. When the battery is full, the icon will appear solid. As the battery charge is used up, the icon becomes empty, and you will need to recharge them soon (see Figure 5-2).

You can also check the battery voltage from the menu system. To do this, select the system icon (the NXT brick—see Figure 5-2). You will see the voltage, and if you are using a rechargeable battery pack you will see an R next to the voltage number (see Figure 5-4).

Just for fun!

The battery icon algorithm works for both rechargeable and disposable batteries, even though they have different fully charged voltages. With batteries, the voltage will slowly creep down through most of the charge, then fall rapidly near the end stage of the charge. Our algorithm monitors this and gives a linear representation of how much battery charge you have left. If the battery charge starts to fall rapidly it usually means the NXT is "running on empty".

So how does leJOS tell if 9V disposable batteries are running low, versus a fully charged battery pack? It turns out the battery pack contains a small nub which presses a small button on the inside of your NXT brick (see Figure 5-5). This allows the battery algorithm to choose the right voltage profile to indicate how much charge remains.

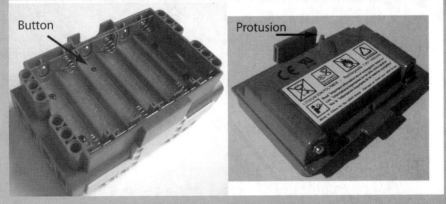

Figure 5-5: The battery pack button

Figure 5-4: Viewing the battery voltage

lejos.nxt.Battery

You can also access the battery charge through the Battery class. The Battery class allows you to determine the voltage produced by the NXT batteries.

```
static float getVoltage()
```

Returns the battery voltage in volts.

```
static int getVoltageMilliVolt()
```

Returns the battery voltage in millivolts.

```
static boolean isRechargeable()
```

Indicates if the battery is rechargeable.

Buttons

All four control buttons on the NXT brick are reprogrammable under leJOS NXJ. You can even use events to listen for button presses and react accordingly when one is activated. This makes it easy to separate the user interface portion of your code in an object-oriented style.

lejos.nxt.Button

The Button class contains static instances of the four buttons (much like the Motor class contains static instances of the motors). These four instances are ENTER, ESCAPE, LEFT, and RIGHT. Often a program needs to wait until a button is pressed. The simplest way to do this is to use the waitForPressAndRelease() method.

```
// Stops code until ENTER pressed
try{
  Button.ENTER.waitForPressAndRelease();
} catch(InterruptedException e) {}
```

You can also use a simple while-loop to stop your code while it waits for the user to press a button.

```
while(!Button.ENTER.isPressed()) {}
```

Java also offers event listeners. Java can initiate an action, or several actions, that are dependent on an event occurring (when a user presses a button, for example). There can be more than one listener waiting for an event to happen. When an event occurs, all the classes that are listening will be notified. The following example shows how to program an event listener.

```
import lejos.nxt.*;
class PlaySound implements ButtonListener {
  public void buttonPressed(Button b) {
    Sound.beepSequence();
  }
  public void buttonReleased(Button b) {}
}
```

This class implements the ButtonListener interface, which contains two method definitions: buttonPressed() and buttonReleased(). All interface methods must be defined in the class implementing the interface. When the button that is registered with this listener is pressed, the NXT will play a series of beeps. Now let's examine a class that registers this listener.

```
import lejos.nxt.*;
class ButtonTest {
  public static void main(String [] args) {
    Button.ENTER.addButtonListener(new
PlaySound());
    while(true){} // Never ending loop
  }
}
```

As you can see in line four, the ENTER button has an instance of the PlaySound listener registered with it (up to four button listeners can be registered for each button). The next line puts our main() method into a never ending loop, but it could just as easily have continued our program. We could also have opted to do this in one class by having ButtonTest implement the ButtonListener interface, and then add itself to the ENTER button.

> **NOTE:** The buttons in leJOS NXJ all make default tones automatically when pressed. If you want to disable the tones so you can use your own sounds, use Button.setKeyClickVolume(). By setting the volume to zero, the tone is disabled:
>
> ```
> Button.setKeyClickVolume(0);
> ```

Speakers

The NXT contains a sound amplifier chip that can play sampled sound through a small speaker. You can even make your own recordings and upload them to the NXT brick. Don't expect this to sound like MP3 quality, however. Because of memory limitations, the NXT can only play low fidelity sound with a low 8-bit sample rate.

lejos.nxt.Sound

The Sound class is responsible for playing sounds. The playTone() method is the most versatile. There are also convenience methods for playing basic sounds.

```
public static void playTone(int aFrequency,
    int aDuration)
```

This method plays a tone, given its frequency and duration. Frequency is audible from about 31 to 2100 Hertz. The duration argument is in hundredths of a second (centiseconds, not milliseconds) and is truncated at 256, so the maximum duration of a tone is 2.56 seconds.

You can also play standard Waveform Audio files (known

as wav or wave) through the speaker. These files must be recorded as 8-bit Pulse-Width Modulation files.

```
public static int playSample(File file, int vol)
```

This method returns an integer indicating the number of milliseconds the file will play for.

There are also several methods for playing unique predetermined sounds.

- public static void beep()
- public static void twoBeeps()
- public static void beepSequence()
- public static void beepSequence()
- public static void buzz()

System Time

Time is reported as the number of milliseconds that have elapsed since the NXT was turned on. Think of this as a time sensor for your robot. This can be useful for timing different actions or time-stamping events for analysis.

- public static long System.currentTimeMillis()

One extraordinary feature of leJOS is the ability to produce time in nanoseconds. A nanosecond is one billionth of a second! Normally you don't need that much timing accuracy, but it can be useful if you are using sensors (such as the light sensor) to record rapid events that require extreme accuracy. You can also use this feature to optimize your code for speed.

- public static long System.nanoTime()

> **NOTE:** Although this method returns numbers to the last billionth of a second, it is actually only accurate to about 100 nanoseconds due to the limitations of the internal timing of the NXT brick. Think of it as an estimate.

Javadocs

The leJOS NXJ API currently has over 450 classes and interfaces, each with a multitude of methods. Of these classes, almost half are standard Java classes (JME or JSE) while the remainder are original classes designed by the leJOS developer team.

This book will attempt to introduce you to the main classes by the leJOS developer team. It would be impractical and tedious to cover every class and every method of every class in this book. If you want to find out more details about a particular class or method, I recommend browsing the Java API docs yourself. They are the most comprehensive resource to find out what classes are included in leJOS NXJ and how to use them.

What are API docs? They are simply a hyperlinked listing of all the classes that you can view with your browser. When you download leJOS NXJ they are included in the installation directory. Simply double-click the index.html file and you will be presented with the main screen (see Figure 5-6).

The complete list of packages are in the upper-left corner. When you click on one, you will see all the classes in that package in the lower-left corner. Click on a class or interface, and you will see all the methods in that class in the main window. If you click on one of those methods, you will see detailed instructions for the method. Get used to clicking on the packages, classes and methods as they will come in very handy as you program your robots.

The leJOS developers periodically modify the API. Please refer to www.lejos.org for the latest API documentation. We include a copy accessible from our website if you want to check out the latest features in case you don't have the latest version of leJOS on your computer. You might even want to bookmark the API docs or create a shortcut link on your desktop, as you will probably access this documentation frequently.

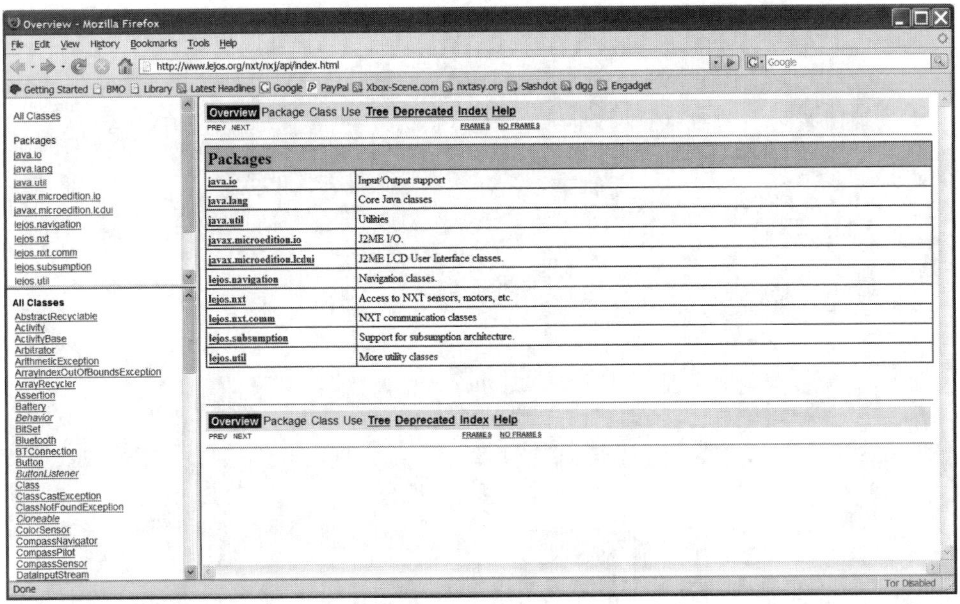

Figure 5-6: Viewing the online leJOS API documentation

CHAPTER 6

An Intro to Navigation

TOPICS IN THIS CHAPTER

- Historical Navigation
- From Ships to Robots

Movement is a central concept in robotics. Often automated tasks require a robot to move from one location to another. We humans take movement for granted because (after the first year of life) it comes very easily to us, requiring little conscious thought. With robots, it is a lot harder.

There are a surprising number of concepts to master in order to perform moves. We will cover each aspect throughout this book, trying out different techniques that will culminate in a robot that is able to perform reliable navigation.

In 2009, the leJOS developers began to notice a number of significant limitations with our existing navigation classes. We spent a lot of time designing a navigational architecture that would be robust, simple, and reliable. To help us define the roles and responsibilities of basic navigation, we looked to past examples of navigation.

Historical Navigation

The British Navy worked out a surprisingly good object oriented design for navigating the seas by assigning specific roles to officers and personnel. In doing so, they formed a chain of command to accomplish reliable navigation. Understanding these roles will help to understand how navigation occurs.

Navigation has been a vexing problem for most of human history. Untold books and movies have depicted the challenge of navigation, such as *The Bounty* (1984), *Longitude* (2000), and *Master and Commander* (2003).

There are two kinds of navigational players: those in the chain of command and those who merely supply intelligence. Those in the chain of command send action commands to other classes. Intelligence actors merely provide data when asked. We'll start at the bottom of the chain and work our way up.

Vehicle

At the very bottom of the chain of command is the vehicle, or in this example, a ship. The ship, like all vehicles, is capable of propulsion and steering. The ship uses wind and sails for propulsion, and a rudder dragging in the water for steering (see Figure 6-1).

Figure 6-1: A sailing ship

There are a diverse number of vehicles in existence: cars, airplanes, boats, hovercraft, and even futuristic walking vehicles. These vehicles use different methods for propulsion and steering.

Although each vehicle is capable of moving, their methods of control are very different. It seems as though we need an expert to control the vehicle. That brings us to the first player on the ship.

Pilot

A ship of the Navy has a wheelman who steers the ship (see Figure 6-2). The wheelman's specialty is that he knows how to control the movement of a ship.

Figure 6-2: The pilot of the ship

Just like there are lots of different vehicles, there are also lots of different pilots: a race car driver, a jet pilot, or even a futuristic battle-mech pilot. Each one of these players only knows how to control a specific vehicle.

Unfortunately, on his own, the pilot gets lost easily because he does not keep track of the position of his vehicle. We need a slightly more intelligent player to aid the pilot.

Navigator

The navigator is a middle-man who is able to take instruction from the captain and then tell the pilot which moves to make in order to get from one location to another (see Figure 6-3).

Figure 6-3: The navigator

The navigator keeps track of position using a coordinate system. This system is a grid containing lines of latitude and longitude. When the navigator is given a target coordinate, he figures out the moves needed to get to the target, and passes those moves to the pilot.

The navigator can only plot the moves if he currently knows where the ship is located. If he has no idea of the current coordinates of the ship, he'll have no idea how to plot a series of moves in order to get to a destination. To obtain the current coordinates, he needs the expertise of several diverse officers who are adept at pinpointing the current location.

Location Providers

On board the ship are a number of officers who are experts at identifying the location of the ship (see Figure 6-4). It is their job to answer the perplexing question, "Where am I?" At their disposal are a number of navigational instruments to help them obtain a location estimate: a sextant to perform sun sighting, a long rope with knots that drags behind the ship to determine speed, a compass, and of course, the stars. Officers can even use a telescope to identify islands in the distance.

Figure 6-4: Locating current position

So what information do these location providers provide? The pertinent information comes down to coordinate information (latitude and longitude) and heading.

Maps

A cartographer uses observational skills to produce a representation of where the ship has been. Often the cartographer was not on the ship, but his maps were (see Figure 6-5). Maps are important to sailing ships because they allow the captain to avoid dangerous reefs and seek out towns. These map features are known as landmarks.

Once a map is created, the information can be used to plot a complicated route that avoids obstacles and seeks out a specific target. The map is used by the...

Path Planner

The navigator merely gets from one point to another, but he is largely ignorant about what sort of obstacles lay between those points. The captain, on the other hand, is a little more sophisticated. Using map data, the captain plots a series of points that avoid dangerous obstacles. The path planner is the captain (see Figure 6-5).

The captain is told by a higher authority where to go, but it is up to the captain to plot the trip so that the crew is not endangered. So who tells the captain what to do?

Figure 6-5: The captain using a map to plot a course

Mission Planning

At the top of the navigational hierarchy are the admirals. They determine where the ship is going and what the crew will do once they get there. Once they know where they want the ship to go, they give the captain a goal, and it is up to the captain to carry out those orders.

From Ships to Robots

Now that you have some idea of the tasks required to make a complete navigational architecture, let's examine how these might be carried out in program code. Some of the goals we had when designing a navigation API are as follows.

1. Compatible with many different types of robots, not just two-wheeled differential steering robots.
2. A general API compatible with NXT and non-NXT hardware.
3. Generate information from a wide variety of different sensors such as compass, GPS, tachometers, gyroscope, accelerometer, range finder—the list goes on and on.
4. Ability to synthesize state information (speed, location, heading) from multiple sensor sources.
5. Ability to use different path finding algorithms.

All of these goals ended up in the Navigation API.

Navigation API

The navigation API provides a convenient set of classes and methods to control a robot. These methods allow the robot to control the direction of movement, move to a point and keep track of coordinates. The next several chapters will guide you through using each of these objects one at a time.

The data flow of these objects is the same as the chain of command described above for the British Navy. All commands flow down the chain of command, and never the other way (although data can flow between different objects). The chain of command is shown in Figure 6-6.

As previously mentioned, the data providers do not issue commands. They merely report location data. Data flow moves to the Pilot, Navigator, or Mission Planner.

One mystery remains: who or what is the mission planner? As it turns out, the mission planner is you! You get to set goals for your robot. Maybe the mission is as simple as bringing a plate of cookies from the kitchen to the living room. Perhaps it is more complex, such as controlling a team of three soccer playing robots. You get to use your intelligence and creativity to plan out your own mission. This book will show you how.

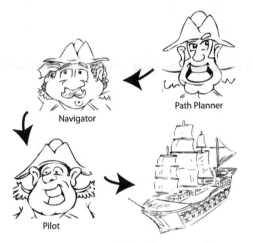

Navigator

Path Planner

Pilot

Figure 6-6: The chain of command

Moves

TOPICS IN THIS CHAPTER

In this section, we will cover the first and arguably most important step in navigation, which is basic movement. The goal is to create vehicles that can perform precise moves. As briefly noted in the previous chapter, the pilot is responsible for driving, steering, and turning a vehicle.

The actual physical characteristics of the robot are hidden from every class in the navigator package except for the pilot. For example, a robot can roll, walk, jump, fly or swim from one location to another, but the external pilot methods to make it move all look the same. The great thing about using pilots is that they allow diverse types of robots to participate in navigation, regardless of their construction.

Metered Moves

As we saw earlier in the book, it is possible to move a robot around using only the motor classes, simply by rotating the motors forward and backward. However, to drive to specific locations, it is necessary to have control over the distances a vehicle moves and the angles that it turns.

One of the primary goals of the navigation classes is to allow precise metered control of vehicle moves. The leJOS developers spent significant time thinking about what moves are and how to represent them with data. Eventually we concluded movement can be represented by four basic types (see Figure 7-1).

1. Straight line travel
2. Arcs (a curved line)
3. On-the-spot rotations
4. Stop

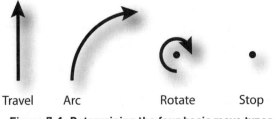

Travel   Arc   Rotate   Stop

Figure 7-1: Determining the four basic move types

By stringing together a sequence of these moves, a vehicle can travel from one location to another. A robot is always performing one of these four moves. Let's decompose these moves further.

A move is composed of two smaller movements—the amount it travels and the amount it rotates throughout the move. These component moves take place simultaneously to create the four complex moves above. Table 7-1 shows each of these moves.

Move Type	Distance	Angle	Permutations
travel	+ or -	0	2
rotate	0	+ or -	2
arc	+ or -	+ or -	4
stop	0	0	1

Table 7-1: Every permutation of distance and angle

Note that distance and angle can be either positive or negative. This means that there are actually two kinds of travel (forward or backward), two kinds of rotate (clockwise or counter-clockwise), and four kinds of arcs. This produces a total of nine discrete moves.

In leJOS, these moves are represented by the Move class, which is a simple container of move information. You can use the Move class to tell a robot what move to make, or you can retrieve a Move object to indicate what kind of move a robot just made. These are the core methods of the Move class:

• getTravelDistance() – the distance moved, normally in centimeters (we will explain this later in the chapter).

- getTurnAngle() – the angle the robot rotated over the course of the move, in degrees.

- getArcRadius() – the radius of the arc that was travelled.

Now that we have examined all the moves a vehicle is capable of executing, let's find out how to actually perform those moves with a physical robot. We will need to call on our reliable friend, the pilot (see Figure 7-2).

The Pilot

There are many types of robots: two-legged walkers, four-legged walkers, car-steering vehicles, and differential robots (such as Carpet Rover from earlier in this book) to name a few. In leJOS jargon, a Pilot is a class that controls a specific robot type. A Pilot can only control one type of robot. For example, the DifferentialPilot class can only control a robot with two drive wheels and a caster wheel for balance.

The DifferentialPilot is the workhorse of the Pilot classes. Most people use this type of robot because differential motor control is capable of all the moves and it is simple to build the chassis.

Figure 7-2: Steering and control courtesy of the pilot

Differential steering has one requirement: the robot must be able to turn within its own footprint. That is, it must be able to change direction without changing x and y coordinates. The robot wheels can be any diameter because these physical parameters are set by the constructor. Once these parameters are set, the pilot class works the same for all differential robots.

The leJOS developers have put in a lot of effort to make the DiffernetialPilot as accurate as possible. The internal equations to keep track of movement are very accurate, down to fractions of an inch. One source of inaccuracy is skidding, which occurs when a robot starts, stops, or performs turns. Depending on the surface, skidding can produce large errors. To combat this, the DifferentialPilot attempts to accelerate slowly and smoothly. You can even adjust the acceleration and speed using these DifferentialPilot methods:

- setAcceleration() – sets the acceleration of the robot in degrees/second^2
- setRotateSpeed() – set the speed at which a robot will rotate, in degrees/second
- setTravelSpeed() – set the speed at which the robot will travel, in units per second

The DifferentialPilot has methods to perform all of the basic moves, which were described earlier in this chapter:

- arc(double radius, double angle)
- backward()
- forward()
- rotate(double angle)
- stop()
- travel(double distance)

Let's try using the DifferentialPilot.

Basic Movement

In this section we will create a project that can perform some basic moves. We'll use the Carpet Rover from chapter 3 (see Figure 7-3). Carpet Rover uses differential steering, thus making it 100% compatible with the DifferentialPilot class.

Figure 7-3: Carpet Rover—a differential steering vehicle

Programming a DifferentialPilot

The DifferentialPilot class requires three basic parameters to work properly:

- Tire Diameter
- Track Width
- Motors

Diameter is the widest measurement from one side of a circle to the other. This is easy to acquire because LEGO prints the diameter right on the tire wall. The tires in the NXT 1.0 kit are 5.6 cm. They produce relatively accurate moves because the rounded surface of the tire almost comes to a point where it meets the floor (see Figure 7-4).

The NXT 2.0 tires are a little smaller at 4.32 cm. The tires are more like street racing tires, with a wide surface contacting the floor (see Figure 7-4).

NXT 1.0   NXT 2.0

Figure 7-4: Comparing NXT 1.0 and 2.0 tires

In order to record accurate rotations when the robot turns, it is important to know the measurement from wheel to wheel, called *track width*. There should be a theoretical point where the wheel touches the ground. The NXT 1.0 tires have a nice point of contact, making it easy to provide an accurate track width (see Figure 7-5).

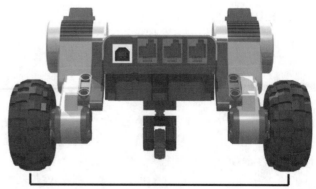

Track Width = 16.4 cm

Figure 7-5: Measuring track width with NXT 1.0 tires

On the other hand, the flatness of the NXT 2.0 tires makes it difficult to determine precise track width. Since LEGO tires are symmetrical, the best way is to measure from the center of one tire to the center of the other (see Figure 7-6). For NXT 1.0 tires, Carpet Rover has a track width of 16.4 cm according to my measurements. For NXT 2.0 tires it is more like 16.5 cm, but measuring closer to the inner points of these tires works better due to the nature of these tires. I used 15.5 cm.

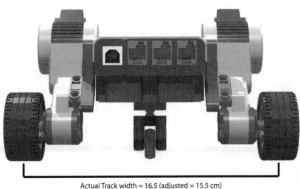

Actual Track width = 16.5 (adjusted = 15.5 cm)

Figure 7-6: Measuring track width of NXT 2.0 tires

The final parameters for the DifferentialPilot are the motors for the left and right wheels. For the differential pilot, these are the motors plugged into the B and C ports.

Now that we know all the parameters of the Carpet Rover, let's lay down some code to test the accuracy of distances and rotations. The following program attempts to drive the robot forward 100 cm, and then rotate three times (360 x 3 = 1080 degrees). We could rotate only once, but three times will allow an average value. Before running the program, try measuring out 100 centimeters on the ground to see how accurately it moves. When it is done moving forward, press the orange button and monitor how accurately it rotates three times.

```
import lejos.nxt.*;
import lejos.robotics.navigation.*;

public class DifferentialTest {

  public static void main(String [] args) {
    //double diam =
DifferentialPilot.WHEEL_SIZE_NXT1;
    double diam = DifferentialPilot.WHEEL_
SIZE_NXT2;
    // double trackwidth = 16.4;
    double trackwidth = 15.5;

    DifferentialPilot robot = new
DifferentialPilot(diam, trackwidth, Motor.B,
Motor.C);
    robot.travel(100);
    Button.ENTER.waitForPressAndRelease();
    robot.rotate(1080);
  }
}
```

Results

Hopefully you have been able to refine the track-width enough so your robot performs three full rotations accurately. It is worth trying the robot out on a few different surfaces to notice how the robot behaves on carpet compared to hard surfaces. Small differences between hard surfaces like linoleum, tile, glass and hardwood also make a difference. Adjust the track-width value for your particular surface.

If you are an owner of the NXT 2.0 kit, you might have noticed that the tires are not entirely consistent from one test to the next. The reason is because the NXT 2.0 tires are wide and flat, therefore different parts of the tire make contact with the surface over the course of a move.

NOTE: Sometimes the motors are reversed, causing your robot to move in directions you did not intend. This is easy to correct:

Just for fun!

Did you know that the DifferentialPilot allows tires with different diameters? If you have access to larger LEGO tires, substitute one of the tires (see Figure 7-7). Then try this alternate constructor out in place of the one used in the code below—but make sure you substitute the correct track-width and wheel diameters:

```
DifferentialPilot robot = new DifferentialPilot(8.16,
4.32, 17.1, Motor.B, Motor.C);
```

Normally you would think that having one tire larger than the other would cause it to drive in a circle, even though it is trying to drive straight. However, watch closely and you'll notice the small wheel turns faster than the large wheel. DifferentialPilot takes the different wheel sizes into account and precisely spins the smaller tire faster, producing a straight line. (Yes, there is some serious math hiding in many of the leJOS classes.)

Figure 7-7: Piloting a robot with different tire diameters

Problem	Solution
Goes backward instead of forward	Set the Boolean parameter to true in the pilot constructor.
Robot turns right instead of left	Swap the motors in the constructor (or swap the port cables)

Carpet Pilot

Now that we have the correct values for the DifferentialPilot, let's try a more substantial program. The short program below produces a simple perpetually moving robot that will move forward until it encounters an obstacle, at which time it backs up and changes direction. For this robot you will need to install an ultrasonic sensor on the robot (see Figure 4-6).

```
import lejos.nxt.*;
import lejos.robotics.navigation.*;

public class CarpetPilot {

  public static void main(String [] args) {
  DifferentialPilot robot = new
DifferentialPilot(4.32, 15.5, Motor.B, Motor.C);
  UltrasonicSensor us = new
UltrasonicSensor(SensorPort.S1);
    robot.forward();
    while(!Button.ESCAPE.isPressed()) {
      if(us.getDistance() < 40) {
        robot.travel(-20);
        robot.rotate(45);
        robot.forward();
      }
    }

  }
}
```

Make sure to substitute the appropriate tire size and track width for your robot and environment.

Move Listeners

One final topic before we move on is the MoveListener interface. The MoveListener is primarily used for communication between the navigation classes, but there are several practical scenarios where you might want to use a MoveListener.

For example, if you have a robot that has an ultrasonic sensor pointed forward and mounted on a motor, you might want it to rotate slightly left or right if the vehicle begins steering around a corner so the sensor is pointed where the robot is traveling. A MoveListener could listen for arc movements and rotate the sensor appropriately.

So how does it work? First, create a class that implements the MoveListener interface. There are two methods your class must implement:

```
moveStarted(Move event, MoveProvider mp)
moveStopped(Move event, MoveProvider mp)
```

Once your class has these implemented, you can add an instance of this class to the pilot, which is a MoveProvider. All pilots are MoveProviders. That is, they make moves and are therefore capable of reporting the moves they just made. The MoveProvider method to add a MoveListener is as follows:

```
addMoveListener(MoveListener listener)
```

Let's alter the example above to include a MoveListener that will output the results of every move to the LCD display:

```
import lejos.nxt.*;
import lejos.robotics.navigation.*;

public class CarpetListener implements
MoveListener {

  public static void main(String [] args) {
```

```
    MoveListener listener = new
CarpetListener();
    DifferentialPilot robot = new
DifferentialPilot(4.32, 16.5, Motor.B,
Motor.C, false);
    robot.addMoveListener(listener);
    UltrasonicSensor us = new
UltrasonicSensor(SensorPort.S1);
    robot.forward();
    while(!Button.ESCAPE.isPressed()) {
      if(us.getDistance() < 40) {
        robot.travel(-20);
        robot.rotate(45);
        robot.forward();
      }
    }
  }

  public void moveStarted(Move move,
MoveProvider mp) {}

  public void moveStopped(Move move,
MoveProvider mp) {
    System.out.println("Moved " + (int)move.
getDistanceTraveled() + " cm");
  }
}
```

There are lots of other scenarios in which you might find the MoveListener interface useful. If you are outputting information to a GUI, you might want to display all the moves the robot made in real-time. You could also have multiple robots reporting moves to a single MoveListener GUI (one path drawn in red and one in blue). In this scenario, the MoveListener code would need to differentiate the robots from each other by checking the MovementProvider variable and drawing the appropriate color.

Just for fun!

The simplest type of Pilot implements only the MoveController interface (see the API docs to explore this interface). This interface allows movement forward and backward only, with no means to rotate. Why did we incorporate such a simple interface? Do such vehicles exist in real life? There are certainly slot cars that only move forward or backward. In the game *Portal 2*, the robot character Wheatley travels all over an enormous research complex on a guided rail system. Of course, the most popular vehicle of this type in wide use is a locomotive. All locomotives have no steering mechanism, and thus only forward or reverse movement is possible.

Try to build a potentially useful MoveController. Imagine a baby plant with high hopes to become a large plant someday. This plant, which happens to be a desert plant, requires a lot of sun. To help that baby achieve its dreams, program your own MoveController that uses a light sensor to find the brightest spot under a window, and that tracks the sunlight throughout the day as the sun moves across the sky. The robot can move the plant forward and backward to keep it in the brightest spot under the window.

CHAPTER 8

Coordinates

TOPICS IN THIS CHAPTER

- Cartesian Coordinates
- Navigator API
- The Rambler

Human language can describe a location in words: "I am in the living room. You are at the corner of 5th Avenue and 3rd Street. She is in Texas." However, these descriptions mean nothing to your LEGO robot because it has no understanding of words. Instead, your robot can understand numbers. In this chapter, we will use a Cartesian *coordinate system* to describe locations.

Cartesian Coordinates

A two-dimensional coordinate system keeps track of two numbers, x and y. Numbers grow larger and smaller along the x and y axes. Both of these axes start at zero and include positive and negative numbers (Figure 8-1). The x and y axes divide the system into four quadrants. Any *point* in a two dimensional area can be plotted on this grid using values of x and y.

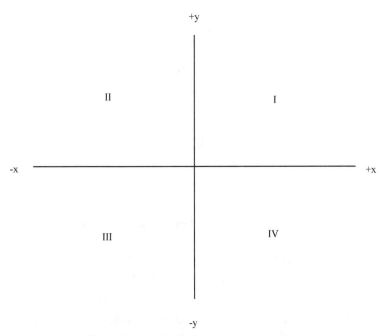

Figure 8–1: A Cartesian coordinate system

Furthermore, in a Cartesian coordinate system, rotations to the left (counter-clockwise) are designated as positive rotations. Perhaps on some other planet in the universe, there is a population that decided to use clockwise rotation as positive. Not this planet, however, so we are stuck with it.

If you rotate +90 degrees, it means you rotate counter-clockwise. Likewise, a rotation of -90 degrees is equivalent to rotating clockwise.

The Navigator API

Now that we know a little bit about coordinates, how do we make a robot drive to a specific location? Earlier in this book we learned about pilots, and how pilots allow a robot to perform moves and drive specific distances. Using a pilot, there is another class called Navigator that tells the pilot how to drive to a specific location (see Figure 8-2).

Figure 8-2: The navigator thinking up some moves

The Navigator accepts a MoveController (any kind of pilot object) in its constructor. It calculates a series of moves to drive from one location to another. The Navigator knows nothing of how the robot works and it does not care how the pilot figures out how to move around. Instead, it just asks the pilot to execute the moves.

The following code (which assumes a pilot exists) instantiates a Navigator and drives to the target coordinate x=100, y=200.

```
Navigator nav = new Navigator(pilot);
nav.goTo(100, 200);
```

It is pretty simple so far. The target coordinates are sometimes known as waypoints. In fact, if you made your own method to generate a series of waypoints, you can feed the coordinates to the navigator using Waypoint objects. Waypoints are used as follows:

```
Waypoint wp = new Waypoint(100, 200);
nav.goTo(wp);
```

Sometimes you need to plot a predetermined path of waypoints for a robot to follow. In this case, you can feed the waypoints to a navigation queue using the addWaypoint() method.

```
for(int x=0, x<1000, x+=100)
    nav.addWaypoint(new Waypoint(x, 20));
```

Rambling Around

To demonstrate coordinate navigation, let's take the trusty Carpet Rover robot from earlier in the book and put it through a randomly generated obstacle course. The following code will make the robot venture from its starting point to some random location and back again. Each time it will display the coordinates so you can see where it is headed. It will wait for you to press the orange button between moves.

```
import lejos.robotics.
navigation.*;
import lejos.nxt.*;
```

Just for fun!

How many ways can you store x and y? You might have noticed leJOS NXJ has a lot of containers that deal with x and y coordinates. We have Point, Pose, Coordinate, Waypoint, and Node. Why so many? Each container serves a specific task, and normally has a set of helper methods that deal with that task. For example, the Node container has methods for adding and removing other nodes from the set (more on nodes later).

```
public class Rambler {

  public static final int AREA_WIDTH = 200;
  public static final int AREA_LENGTH = 200;

  public static void main(String[] args)
throws Exception {

    DifferentialPilot p = new
DifferentialPilot(DifferentialPilot.WHEEL_
SIZE_NXT2, 15.5, Motor.B, Motor.C);
    Navigator nav = new Navigator(p);

    // Repeatedly drive to random points:
    while(!Button.ESCAPE.isPressed()) {
      System.out.println("Target: ");
      double x_targ = Math.random() * AREA_
WIDTH;
      double y_targ = Math.random() * AREA_
LENGTH;
      System.out.println("X: " + (int)x_targ);
      System.out.println("Y: " + (int)y_targ);
      System.out.println("Press ENTER key");
      Button.ENTER.waitForPressAndRelease();

      nav.addWaypoint(new Waypoint(x_targ, y_
targ));
      nav.addWaypoint(new Waypoint(0, 0, 0));
    }
  }
}
```

Make sure to change the wheel size and track width to the appropriate sizes for your robot. Be sure your motors are plugged into ports B and C.

Testing and Results

You need an area of about four meters square for this course. If you have more or less than that, you can change the two area constants at the start of the program (these values are in centimeters). Try placing a dime at the starting location so you can see how much accuracy it loses each time it ventures out. It is important to have Carpet

Rover precisely lined up along the imaginary x-axis at the start. You could even attempt to place a dime at the randomly generated target each time.

Carpet Rover does a good job of measuring distances but the robot has a weakness when it rotates. You will probably notice that after the first waypoint the accuracy is off slightly, which causes it to miss the second waypoint even further, and so on.

The errors are cumulative, so the longer your robot travels, the further off course it becomes until the location coordinates are essentially meaningless. This is a chronic problem with tachometers. You can minimize drift problems by running the robot on a smooth, hardwood floor, but you can also seek to improve rotation accuracy. This book will attempt to make navigation more accurate in upcoming chapters.

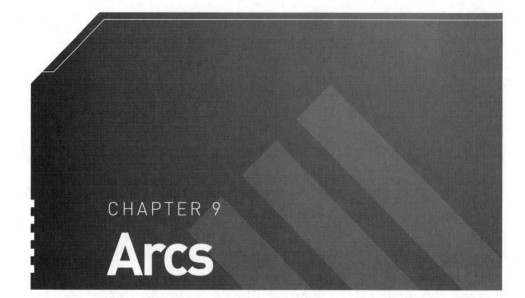

CHAPTER 9

Arcs

TOPICS IN THIS CHAPTER

 Arcs

 Navigation with Steering

The most popular robot chassis is the differential steering robot. A version of this chassis, Carpet Rover, was introduced earlier in this book. The reason for its popularity is because these robots are easy to build and they are capable of performing a diverse set of moves.

However, the most popular type of vehicle on earth, the automobile, uses an analog steering mechanism. This method of travel is ideal for humans because steering cars are stable at high speeds and they are easy to operate.

It would be a big omission if the navigation package left out the most popular vehicle on the planet. However, steering poses a special challenge because a steering vehicle is unable to rotate within its footprint. Instead, it can only change direction by driving an arc.

Arcs

When driving a car, you rotate the steering wheel to make a turn. This causes the vehicle to drive an arc. If you turn the wheel slightly, it causes a shallow arc and if you turn the wheel sharply, it causes a tight arc (see Figure 9-1).

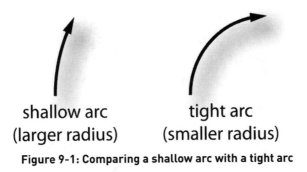

shallow arc
(larger radius)

tight arc
(smaller radius)

Figure 9-1: Comparing a shallow arc with a tight arc

What is an arc and how can we tell a robot how to drive a specific arc? Several components define an arc, notably the radius and angle (see Figure 9-2). The arc angle (degrees) can also be expressed as a length in units, such as centimeters. An arc with a larger radius will produce a shallow arc, and an arc with a smaller radius will produce a tight arc (see Figure 9-1).

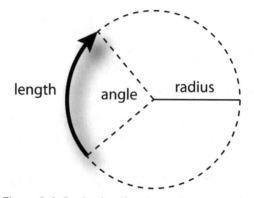

Figure 9-2: Reviewing the components of an arc

In chapter 7, we examined four types of moves: straight travel, arc, rotate, and stop (see Figure 7-1). We also noted that within the arc move, there are actually four sub-types. These types are determined by the direction of movement around the arc, and whether the center of the circle lies to the left or right of the robot when it is performing an arc (see Figure 9-3).

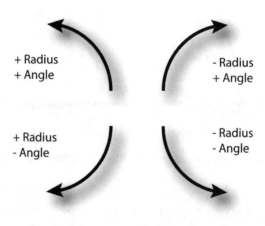

Figure 9-3: The four kinds of arc moves

Note: This is sure to confuse you even more. Recall earlier that in a Cartesian coordinate system, positive angles are counter–clockwise. Therefore, technically the right side of Figure 9-3 is incorrect. When a robot travels forward and right along an arc, the angle should be negative. However, we used the opposite approach with leJOS because it is confusing to give commands to a pilot class using a true Cartesian system. In leJOS, a positive angle moves the robot forward while a negative angle moves it backward.

For example, if you ask a robot to travel an arc of 110 degrees and a radius of -25, it will drive forwards and to the right. If we ask it to arc 110 degrees with a radius of 25, it will arc to the left. If we make the radius larger, such as 50, the arc will not be as tight. Now that we are able to describe an arc with numbers, let's see how we can navigate with arcs.

Analog Steering

Any class that implements the ArcMove-Controller interface is capable of driving an arc. The principle method in this interface is the arc() method, which requires the two parameters described above: radius and angle.

Several pilot classes implement the Arc-MoveController interface, such as DifferentialPilot and SteeringPilot. The method of movement is irrelevant to the API. Typically a wheeled robot is performing movement, but to the API, it does not matter if it is a walking robot or even a hovercraft. As long as the vehicle can fulfill the basic movements, it is acceptable to the API.

Steering robots are a little more finicky than differential robots because the steering requires calibration. The next

Just for fun!

All of the four moves (travel, arc, rotate and stop) can be represented in mathematical terms by an arc. To travel a straight line, the radius size is infinity. To perform an arc, the radius is greater than zero and less than infinity. And to rotate on the spot, a radius of zero is used (reducing the arc to a point).

chapter contains instructions to build and program a steering robot, but for this chapter, we just want to explore navigation with arcs, so the trusty Carpet Rover is used instead.

We can force the Carpet Rover to emulate an analog steering vehicle. We merely need to increase the minimum turning radius of the DifferentialPilot from zero to a positive number using the setMinRadius() method.

```
myPilot.setMinRadius(34);
```

Now if you pass the DifferentialPilot object to a Navigator, it will treat the robot as though it has analog steering capable of turning with a radius of 34 centimeters.

Steering and Navigation

The previous chapter demonstrated differential steering navigation. To move from one coordinate to another, a differential robot simply rotates in place until it points at the destination coordinate, then it drives the appropriate distance (see Figure 9-4a).

Steering robots do not have the luxury of turning in place, so the navigator must calculate a path using arcs and straight travel to move to the destination. The algorithms are contained in ArcAlgorithms, a special class just for performing these calculations. Moves are plotted by driving an arc until it points at the target, followed by a straight-line travel to the target (see Figure 9-4b).

In the previous chapter we constructed a pilot and then gave it to the Navigator class as a parameter in the constructor. The same constructor applies to steering vehicles. The following few lines of code create a Navigator object and cause it to move to the point in Figure 9-4.

```
Navigator nav = new Navigator(arcPilot);
Nav.goTo(40, 50);
```

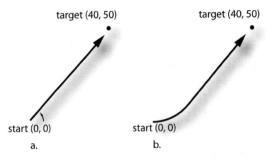

Figure 9-4: Comparing differential and analog steering

Heading

The pose of a robot represents a position of a robot at any given time (in the past, at present, or in the future). The pose of a robot includes coordinates and the heading of the robot (the direction it is pointing). There are two classes in the navigation package that contain heading data: the Pose class and the Waypoint class (more on these later).

With steering vehicles, such as automobiles, the final pose of the vehicle is critical. When you park a car in a driveway, you cannot leave the vehicle crossways or diagonally, otherwise your neighbors will worry about you. It is especially important to line up the vehicle before attempting to enter a garage, otherwise the vehicle could hit the entryway. The final pose of the vehicle is important.

Using the Navigator class, you can specify the final pose of the vehicle as follows.

```
nav.goTo(40, 50, 90);
```

The first two numbers in this method specify the Cartesian coordinates, while the number 90 indicates the final Cartesian angle.

You can also indicate the final pose using a Waypoint object, as follows.

```
Waypoint target = new Waypoint(40, 50, 90);
nav.goTo(target);
```

Putting it all together

Let's create a full program to navigate to several points in coordinate space. This will help us to see some of the nuances with steering navigation.

```
import lejos.nxt.*;
import lejos.robotics.navigation.*;

public class Steering {

  public static void main(String[] args) {
    // Make sure to use correct tire size and
track-width!
    ArcMoveController pilot = new
DifferentialPilot(5.6, 16.4, Motor.B,
Motor.C);
    pilot.setMinRadius(34);
    Navigator nav = new Navigator(pilot);
    nav.goTo(40, 50, 90);
    Button.ENTER.waitForPressAndRelease();
    nav.goTo(0, 0, 0);
    Button.ENTER.waitForPressAndRelease();
    nav.goTo(40, 50, 180);
  }
}
```

NOTE: Make sure to use the correct tire size and track-width in the above code. If your accuracy does not seem very good, you probably need to adjust one of these two numbers in the DifferentialPilot constructor.

When you run this code, it will attempt to get to the target of 40, 50 and a heading of 90 degrees (see Figure 9-5a). The path it travels is predictable and expected. When you press the orange button it will drive back to the original starting point.

The second time you press the orange button, it will drive to the same coordinate but it will arrive with a heading of 180 degrees (Figure 9-5b). Surprised at the convoluted path it took to get there? This path was generated by the ArcAlgorithms class. It calculates all the different path permutations and chooses the shortest path possible to get to the destination. As it turns out, the solution it found involves a lot of reverse arc traveling.

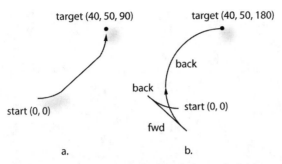

Figure 9-5: Final heading affects the path driven

One other wrinkle with the navigation algorithms is that the robot will always drive forward in between two arc movements. It will never travel backwards for the straight portion of the path. This leads to some seemingly odd behavior. In the code above, try substituting the first destination with this code.

```
//nav.goTo(40, 50, 90);
nav.goTo(20, 0, 0);
```

As expected, the first move just drives forward 20 centimeters. However, to get back to the start, you would expect it to back-up 20 centimeters. This cannot happen, however. Remember, it can only drive forward. Instead, the robot executes a complicated move to get there.

The reason it will never drive backwards for the straight travel (which can be the longest part of the move) is because we felt vehicles should drive forward for long stretches at a time, especially given that most sensors will point forward with actual robots.

You are now familiar with arc navigation. The next chapter contains a project using a sophisticated steering robot to perform navigation.

CHAPTER 10

Ackerman Steering

TOPICS IN THIS CHAPTER

In the previous chapter, we used a differential steering robot to perform arc movements, which simulated the behavior of a steering vehicle. This chapter contains instructions to build and control a true steering vehicle. Like the previous example, this vehicle is capable of full coordinate navigation. Let's find out how to design and program a proper steering vehicle.

Differential Drive

Steering vehicles seem simple on the surface. After all, it looks like it just needs two wheels to drive the vehicle and two to steer. In fact, there are a few complexities if you want to do it right.

The first complication has to do with the rear drive wheels. When two wheels exist on the same axle, things work fine as long as the vehicle drives straight. However, if you try to steer, the inner wheel will drive a smaller circle than the outer wheel (see Figure 10-1).

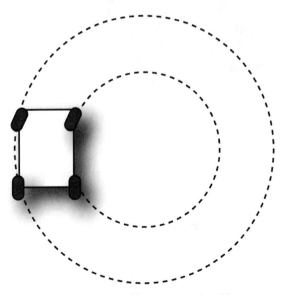

Figure 10-1: Different wheel radii

Why does this pose a problem? The reason is because the wheels are both on the same axle. As you can see in figure 10-1, the inner wheel travels a shorter distance than the outer wheel. The axle can only turn at one speed, therefore the axle will either break (unlikely) or the wheels will skid on the surface, wearing out the rubber tires and causing wider turns.

Normally a differential is used to allow both tires to turn at different rates. However, since the NXT kit does not include a differential gear train, we cannot use this solution. Instead, we will remove this problem by using one rear wheel instead of two. The vehicle in this chapter uses a single rear wheel for simplicity (see Figure 10-2).

Figure 10-2: Driving with a single rear wheel

Ackerman Steering

The second problem is similar to the first, except it has to do with the steering mechanism. When the vehicle is making a turn, normally the inner wheel drives a smaller circle than the outer wheel (see Figure 10-1). You might not think this is a problem, because the steering wheels are not connected by a single axle. However, if both tires are angled in the same direction, they will both try to steer the same sized circle.

As you can see in Figure 10-1, the inner steering wheel travels a smaller circle than the outer wheel, so it should be turned at a sharper angle to compensate. If the vehicle does not compensate for this, lateral skidding will occur and the vehicle will not turn as sharply or accurately.

The solution is to use Ackerman steering. This mechanism allows the axle of each wheel to point directly at the center of the turning circle, producing tighter turns with less slippage (see Figure 10-3).

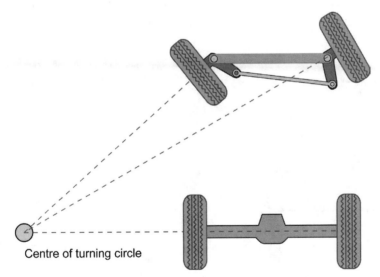

Centre of turning circle

Figure 10-3: The inner wheel steering sharply

Steering Robot

Let's try building an analog steering vehicle. This vehicle is called Ackerbot in honor of the inventor of the steering mechanism.

1

15

2x

2x

1x

2x

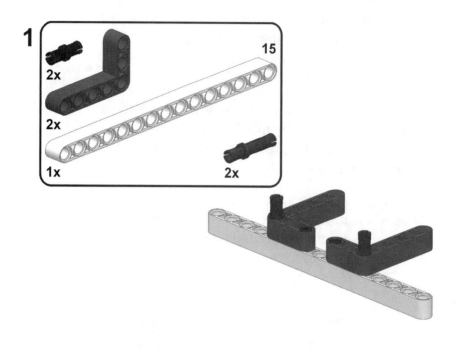

2

11

2x

4x

3

2x

2x

2x

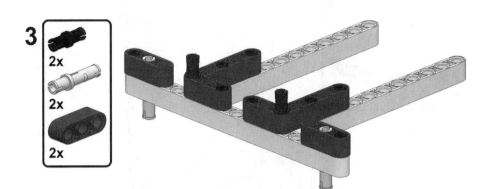

4   9

1x   4x

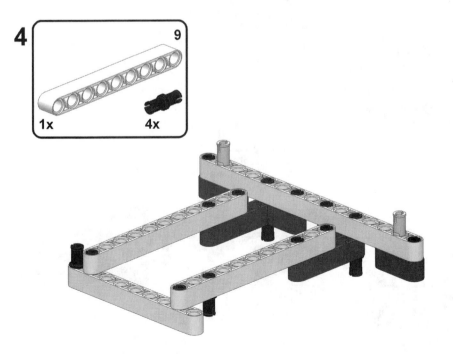

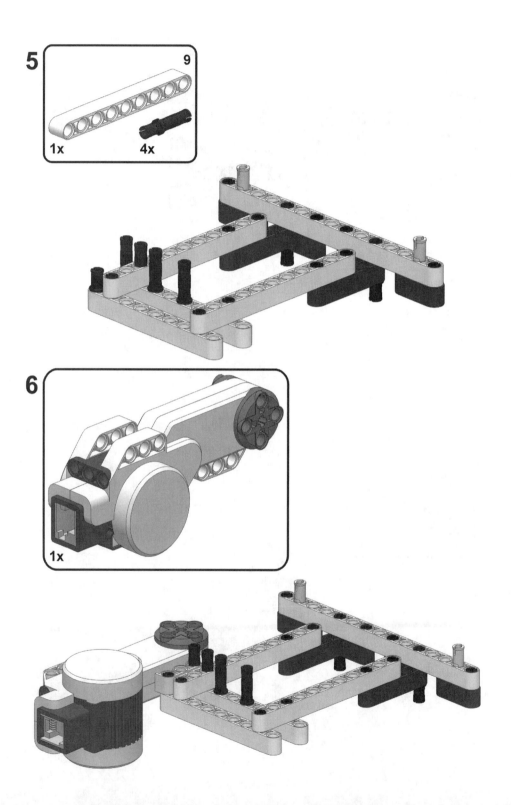

7

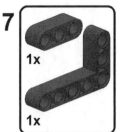

1x

1x

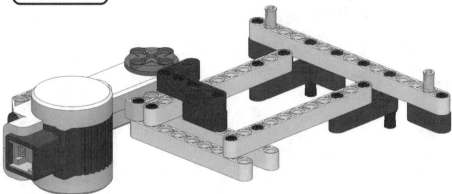

8

9

2x

2x

2x

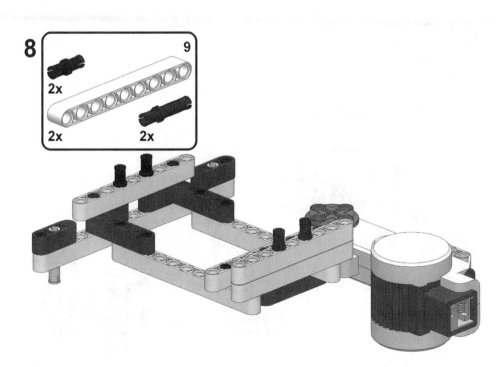

9

1x   2x

10

3

1x

1x   2x

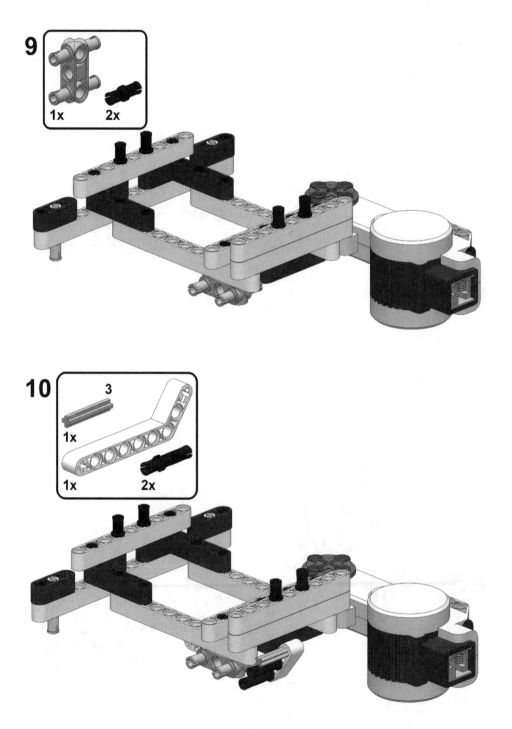

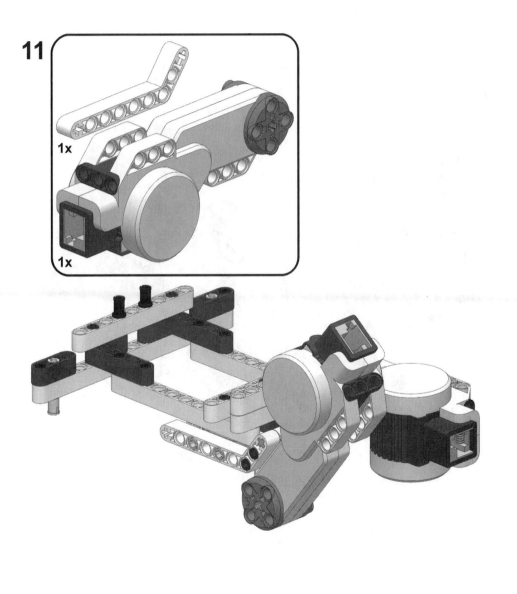

12

7

1x   2x

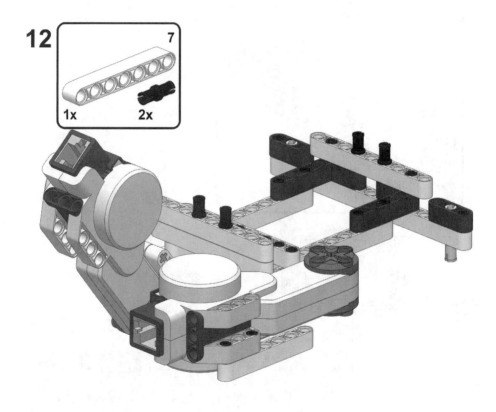

13

2x

1x

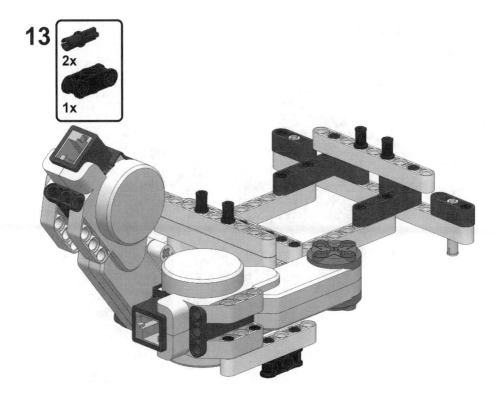

14

6

1x   1x

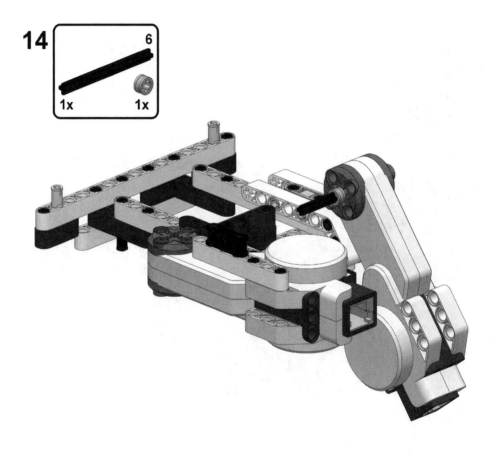

15

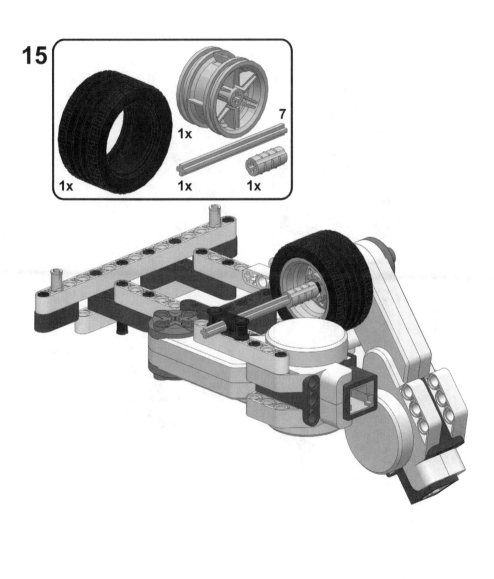

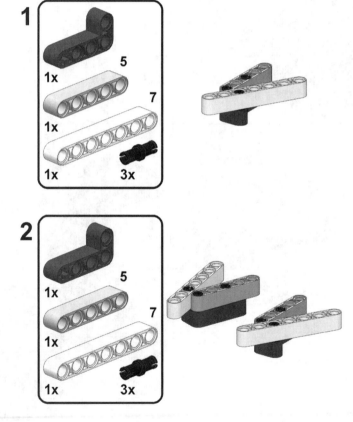

3

4x

5
2x

4

7
2x

5

11
1x   2x

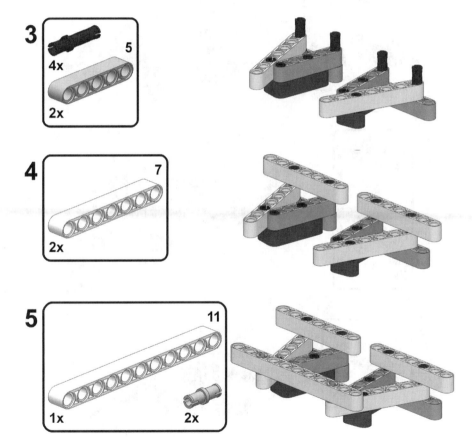

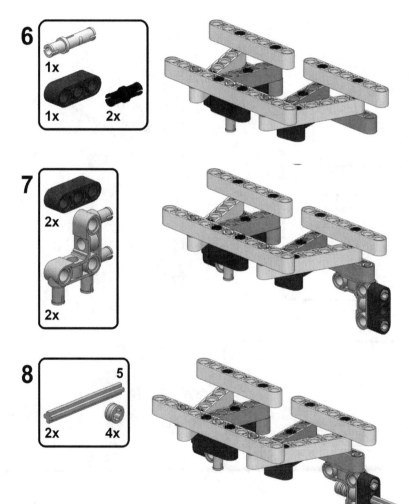

6
1x
1x 2x

7
2x
2x

8
5
2x 4x

9

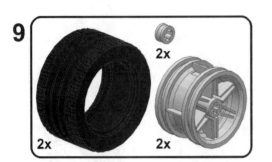

2x

2x

2x

16

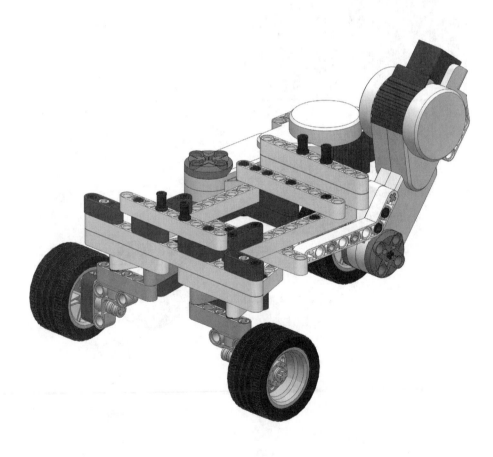

17

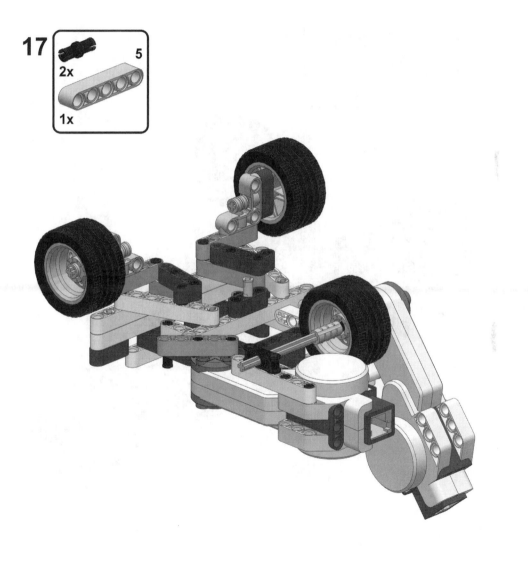

18

9

1x   1x

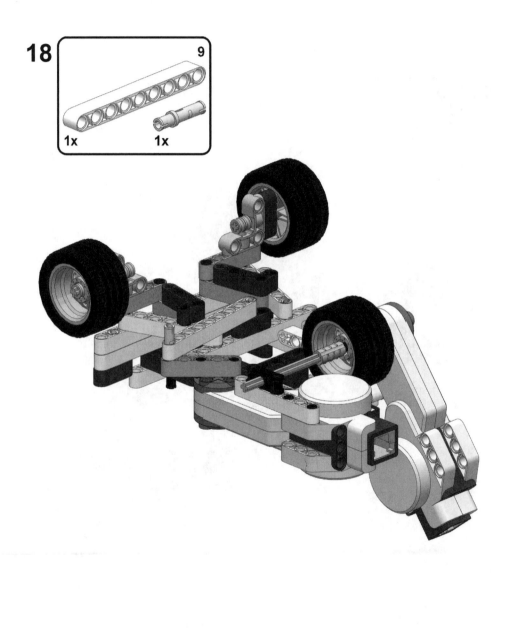

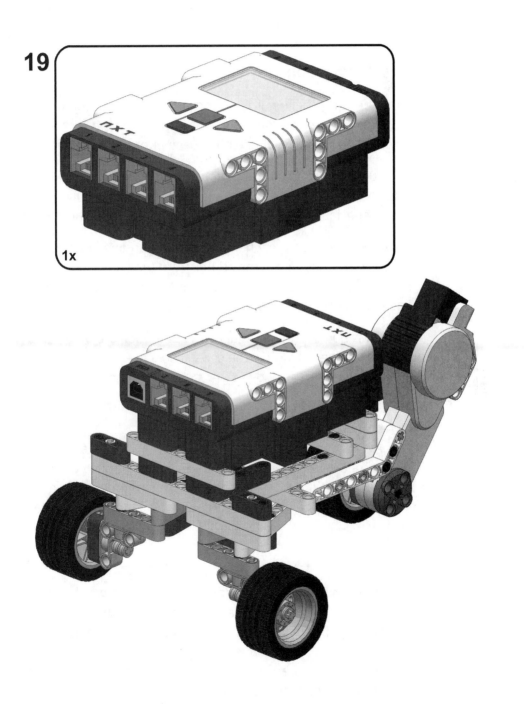

Connect the slanted drive motor to port B with a medium cable. Connect the steering motor to port C with another medium cable.

Auto-Calibration

With steering vehicles, it is important for the wheels to point straight ahead when the vehicle attempts to drive forward. If they do not, the robot will be off course after only a few moves.

You could manually align the steering using your eye, but it is faster and more accurate to have the NXT perform alignment using a specialized calibration routine involving the tachometers.

This routine rotates the wheels all the way in one direction until the motor encounters resistance from the steering frame. It then records the tachometer reading. Now it knows how much it needs to turn the motors to turn all the way in one direction.

The calibration routine then does the same for the other direction. It also needs to know the center, so it averages the two tachometer readings to find out the value in the immediate center.

This calibration routine is part of the SteeringPilot class, in the method calibrateSteering(). However, since our robot uses Ackerman steering with an asymmetrical steering arm, this method does not work. Instead, we will have to write our own custom calibration method, using a similar methodology. You can view the calibration method in the code sample below.

SteeringPilot

Ackerbot is technically capable of steering a whole range of arcs, with the smallest arc radius of about 32 cm (the minimum radius). However, the SteeringPilot class is only capable of doing either a left arc or a right arc—no middle ground. It cannot perform a shallow arc, only minimum radius arc (the tightest arc).

The reason the SteeringPilot is incapable of a range of arcs is because it is too difficult to calculate the entire range of steering radius given the steering angle. Therefore the code for the steering robot performs only three steering angles: left, straight, right.

Adjustments

Because we are using standard LEGO kits with almost identical parts, your robot should drive with about as much accuracy as mine. If you want to try honing the steering to make your robot even more accurate, you can test out some patterns to see how it performs.

First, let's test the turn radius. The following code calibrates the steering, then drives the robot in several 360 degree circles:

```
import lejos.nxt.*;
import lejos.robotics.RegulatedMotor;
import lejos.robotics.navigation.*;

public class Ackerbot {

  static RegulatedMotor motor;
  static int left = 0;
  static int right = 0;
  static int center = 0;

  public static void main(String[] args)
throws Exception {
    Ackerbot.calibrate(MotorPort.C);
    Ackerbot.recenter(Motor.C);
    Motor.C.setAcceleration(200);

    double MINTURN_RADIUS = 31.75;
    SteeringPilot p = new
SteeringPilot(SteeringPilot.WHEEL_SIZE_NXT2,
Motor.B,
      Motor.C, MINTURN_RADIUS, 48, -42);

    p.arc(MINTURN_RADIUS, 360);
    Button.ENTER.waitForPressAndRelease();
    p.arc(-MINTURN_RADIUS, 360);
```

```
    }

    public static void calibrate(MotorPort
port) {.
      NXTMotor m = new NXTMotor(port);
      m.setPower(20);
      m.backward();
      int old = -999999;
      while(m.getTachoCount() != old) {
        old = m.getTachoCount();
        try {
          Thread.sleep(500);
        } catch (InterruptedException e) {
        }
      }

      right = m.getTachoCount();
      center = (59 + right);
      left = center + 59;
    }

    public static void recenter(RegulatedMotor
steering) {
                    motor = steering;
                    motor.setSpeed(100);
                    motor.rotateTo(center);
                    motor.flt();
                    motor.resetTachoCount();
      }
    }
```

Just for fun!

There is one other analog steering vehicle in this book (Chapter 19). It was built primarily as a remote control vehicle, but try converting it to an autonomous robot by adapting the code in this chapter.

As you can see in the code, the starting minimum turning radius of this robot is 31.75. Set it down, start the program and watch Ackerbot calibrate the steering. Once complete it will drive two circles. Press the orange button after the first circle.

Ideally, you want Ackerbot to drive a full circle and end precisely where it started. If you robot falls short, try increasing the MIN_RADIUS value until it does a complete lap. Once you have the MIN_RA-

DIUS value properly adjusted, try measuring the radius of the circle it is driving (radius is diameter divided by two). Does the measured radius in fact end up the same as MIN_RADIUS?

Now that the SteeringPilot constructor parameters are honed, let's try using it with a Navigator. Comment out the code in the main() method that makes it drive two circles. Replace this code with the following code to see how well it navigates.

```
Navigator nav = new Navigator(p);
nav.goTo(40, 50, 90);
nav.goTo(0, 0, 0);
```

CHAPTER 11

Localization

TOPICS IN THIS CHAPTER

Localization is the ability to determine where you are. Engineers and inventors have come up with a number of localization methods since the dawn of time (see Figure 11-1). As mentioned earlier, the most serious early efforts were to estimate the correct position of a ship at sea with very few visual landmarks—perhaps the occasional island. This book will examine several methods of localization. First, let's find out more about how people determine where they are.

Figure 11-1: Locating position

Localization

Earlier, this book introduced the concept of a coordinate system, which is capable of indicating the x and y position of a robot. Using coordinates, we were able to tell the robot to move to a specific position. But how do we find the current position of the robot?

Localization can be determined using two main strategies: odometry and landmark navigation. Imagine yourself in a very large and empty gymnasium. There are no visual marks on the floor, except for two points marked A and B at opposite ends of the room (see Figure 11-2).

Figure 11-2: Traversing an environment with limited data

Now imagine that you are placed on point A, blindfolded, then told to make your way to point B. To navigate, all you would have at your disposal is *proprioception*, the ability to feel the orientation of your legs, which would allow you to estimate the length of each step. This is a biological form of odometry. By counting the number of steps you make, you would be able to roughly indicate when you were near point B.

As you might guess, the estimate when you arrived at point B is likely to be somewhat off the actual mark. In fact, try it right now in your living room. Place a tiny piece of paper at the far end of the room, place your hand over your eyes, and try to walk until you are right on the mark (make sure not to feel for the paper with your toes, and try not to use audio landmarks such as a noisy air vent or open window). Over a distance of about 20 feet you'll probably be off about a foot.

The second way to navigate is by using landmarks. Imagine the same gymnasium, except that it has a grid pattern of poles spaced five feet apart (see Figure 11-3). If you note beforehand that point B is four poles north and two poles east, you can then close your eyes and feel your way along the poles until you get to the destination.

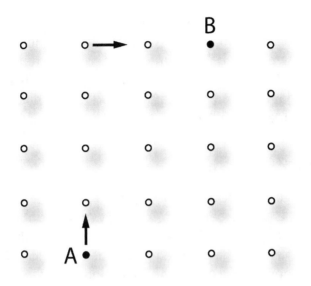

Figure 11-3: Traversing an environment with fixed landmarks

So which navigation system is more reliable? As you might guess, landmark navigation is by far the more reliable method. If you use blind odometry to get from point A to point B, and then keep repeating this back and forth without ever peeking, you will be hopelessly lost after a few rounds. However, you could navigate around the poles for hours and still maintain your location, as long as you accurately note how many poles you have traversed and in which direction.

Landmark navigation works consistently for long periods of time. In the gymnasium example above, instead of poles you could have performed landmark navigation merely by keeping your eyes open and walking to the B on the floor. In fact, you can navigate using many different kinds of landmarks, including audio (a radio), visual (a light house), smell (a rose bush), temperature (a heat vent) and radio waves (GPS satellites).

You use landmark navigation every time you walk to the fridge. Humans use fixed landmark navigation by looking at local landmarks, such as your neighborhood supermarket, a tree, street signs, or any number of notable objects. GPS uses this method by noting a position relative to four or more satellites.

Landmark navigation doesn't worry about where you have been, just where you are at a particular moment. Odometry, on the other hand, estimates where you are based on your previous position. We will experiment with fixed landmark navigation and GPS later in this book, but for now let's examine odometry.

Odometry

The most basic localization method is odometry. Using odometry, a robot keeps track of every movement it makes starting from a point of origin (see Figure 11-4). Normally odometry relies on tachometers to keep track of wheel or leg movement—a form of proprioception.

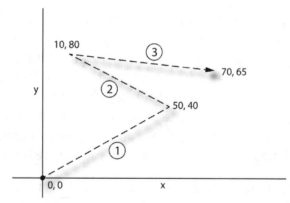

Figure 11-4: Using odometry to estimate coordinates

It is important to stress that odometry is an estimate of the current position. It cannot tell with certainty an exact location.

Humans have used a similar process called *dead reckoning* (also sometimes called orienteering) for a long time. In fact, this method of navigation is used to some degree by all animals, including humans. The art of dead reckoning has been refined by sailors, pilots, geologists, forest rangers, and hikers. The only information needed for dead reckoning is direction and distance.

Direction is usually obtained using a magnetic compass. However, direction can also be determined by recording the rotation of wheels using a tachometer. For example, when one wheel rotates forward and one rotates backward, the robot rotates. By measuring the amount each wheel rotated we can approximate the angle it rotated. Likewise, distance can be approximated by keeping an accurate record of wheel rotations while travelling.

Just for fun!

Do you know the difference between direction and heading? They are in fact two different things. Direction is the direction your face is pointing. If you are looking at the North Star, your direction is north. However, if you sidestep to the east, your heading is east, even though you are still facing north (see Figure 11-5). Heading does not occur when you are standing still, only when you are moving. GPS is only capable of providing heading. If you try to get direction from a GPS while standing still, the direction value is more or less random.

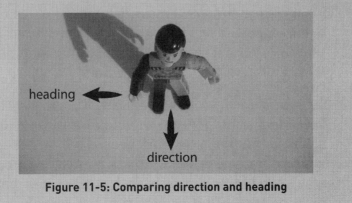

Figure 11-5: Comparing direction and heading

There are some advantages and disadvantages to using odometry with a robot. With smooth wheels, a flat floor, and a good tachometer, a robot will outperform a human in estimating the distance it has traveled. The accuracy of calculating and storing coordinate points is also better in robots than in humans, although a pad of paper often helps humans to keep track of previous travel.

There is a downside to robots, however. Most robots are unable to self-correct their course by analyzing a situation. The robot in this chapter has no ability to visually recognize a target or landmark (we will attempt this later in the book using a technique known as Monte Carlo Localization).

To update coordinates, the robot must perform trigonometry calculations. Luckily for us, leJOS NXJ comes with navigation classes to perform the math.

OdometryPoseProvider

Odometry calculations are performed in the OdometryPoseProvider class. This class updates coordinate data using geometry.

The OdometryPoseProvider class requires a MoveProvider in the constructor. Internally, the MoveProvider updates the OdometryPoseProvider every time a move is made. By updating it with all the latest move data, the class can keep track of the estimated location. All of the Pilot classes implement the MoveProvider interface. That is, when they make a move, they can report the move data via a Move object.

A MoveProvider does not necessarily need to be a pilot. Imagine if a programmer named Gerald created a class called GeraldPoseProvider. Every time he starts or stops a move, he types the information into the computer. This may sound strange to have a MoveProvider that relies on human input, but it would be a true MoveProvider which could be used in the OdometryPoseProvider constructor to estimate Gerald's pose, as long as he accurately provided updates on his movements.

Let's examine an example that moves a robot around while keeping track of the pose using an OdometryPoseProvider. Once again, we'll use the workhorse of this book, the ever reliable Carper Rover.

```
import lejos.nxt.*;
import
lejos.robotics.localization.OdometryPoseProvider;
import lejos.robotics.navigation.
DifferentialPilot;

public class Odometry {

  public static void main(String[] args) {
    double diam =
DifferentialPilot.WHEEL_SIZE_NXT1;
    double width = 16.45;

    DifferentialPilot robot = new
DifferentialPilot(diam, width, Motor.B, Motor.C);
    OdometryPoseProvider pp = new
OdometryPoseProvider(robot);

    robot.rotate(90);
    robot.travel(100);
    robot.arc(30, 90);
    robot.travel(50);

    System.out.println("End: " + pp.getPose());
    Button.waitForPress();
  }
}
```

After you run this program, it should perform a series of moves. When complete, it will output the following approximate final pose.

- x = -81 cm

- y = 130 cm

- heading = 180 degrees

Notice this example does not use the Navigator class? It

only uses DifferentialPilot, yet it can report the coordinates after each move. Localization like this could be useful for a remote control car that uses odometry to keep track of its coordinates. You could try navigating this car blindly through your house using a map, even if it is in another room, by having it report back the coordinates.

In earlier chapters, we created a Navigator class. What you probably did not realize is that it automatically created an OdometryPoseProvider internally. You can access this by using the Navigator.getPoseProvider() method, which in turn allows you to check the current pose. There is also an alternate Navigator constructor which accepts a PosePro-vider as a parameter, allowing you to specify the type of PoseProvider to use.

Odometry Error

The leJOS developers have tried to eliminate as many error sources as possible from the navigator package, but some error will always remain. It is a good idea to be aware of the source of errors so you can do your best to eliminate them.

- Inaccurate measurements of the track width (distance between the two wheels) can result in inaccurate turns. This can be especially hard to measure when the contact area of the floor and the tire is wide (see Figure 7-4).

- Surface texture of the floor can impact accuracy. On a small scale, the ground can have unevenness, such as the spacing between tiles, the spaces between hard-wood floor slats, or lumpiness of carpets. This can af-fect rotation and distance characteristics significantly. Try measuring the same movement between carpet and hardwood and you will see a noticeable difference.

- Wheel slippage can occur when a robot is starting and stopping. This is partly a function of surface texture and tire grip. It also has to do with how fast a robot accelerates. You can minimize this effect by lowering the acceleration speed of the pilot using the setAcceleration() method.

- Backlash is a problem that occurs due to the tiny spaces between gears. Even the internal gearing of NXT motors contain backlash. You can feel the backlash by lightly applying torque to the motor axle with your fingers. The backlash is a dead-spot where the axle turns freely without any resistance.

- Differences between tires can have a surprising effect on accuracy. This is most obvious when you build two robots and attempt to drive them through the same path. Variation caused by differences in the tires can add up to small variations in the accuracy of the robot. Trying different combinations of tires can help minimize this effect.

- Weight distribution of the robot can also have an effect. When the robot rotates in one spot, the weight distribution might be more towards the front or back of the robot. If too much weight is placed on the castor wheel, you can get rotational wobble.

- Robot construction issues can affect accuracy. If one or more wheels on the robot are not level and driving straight ahead, it can alter the robot course. Saggy wheel supports or wobbling wheels can do the same.

These errors will always plague dead reckoning, which makes landmark navigation more appealing. Keep reading through this book to try out more reliable methods of localization.

CHAPTER 12

NXJ Control Center

TOPICS IN THIS CHAPTER

- File Browser
- Settings
- Sensor Monitor
- Motor Control
- Data Logger
- Remote Console

The leJOS platform includes a number of useful utilities to aid robotic development. These utilities are located in the bin directory of your leJOS NXJ install. There are over a dozen total, but we will cover the core utility, called NXJ Control Center.

NXJ Control Center

NXJ Control Center is a comprehensive utility containing many of the other utilities in the bin directory. This mature user interface allows you to control motors, read sensors, change NXT settings and more, all from a PC. Running the utility is simple, assuming you installed leJOS and Java. Simply double click nxjcontrol.bat.

The main screen has options to connect to an NXT brick (see Figure 12-1). Turn on your brick and—depending on how you want to connect—plug it into the USB port or make sure Bluetooth is enabled. Click the search button and the program will search for all available NXT bricks. If you want, you can narrow the search by typing a name or selecting only USB or Bluetooth. Select your brick from the list and click Connect.

The main screen contains three sections selectable by radio buttons: LCD, RConsole, and Data Log. LCP has all the functions available to the LEGO Communications Protocol. The next section contains the Remote Console for debugging output. The final selection allows data capture on your PC via the DataLog class. We'll cover these later, but first let's examine the different tabs available to LCP.

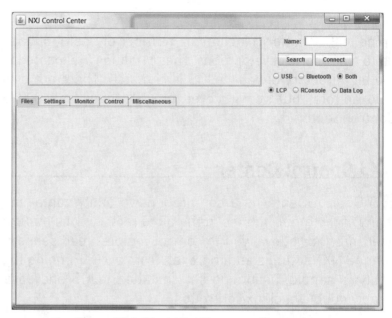

Figure 12-1: Displaying the connection screen

NXJ File Browser

The first tab contains the file browser. This is the main utility for file management. It's easy to compile and upload code, but often you want to upload data files to the NXT brick, such as images, sounds, or map data.

> **NOTE:** File browsing is available through a standalone application called NXJ File Browser. To run the standalone utility, double-click the nxjbrowse.bat file in the bin directory.

The file browser displays the contents of your NXT brick (see Figure 12-2). There are buttons at the bottom of the screen to delete files, upload files to the NXT brick, and even download files to your PC. To delete a file, first place a checkmark next to it before hitting the delete button.

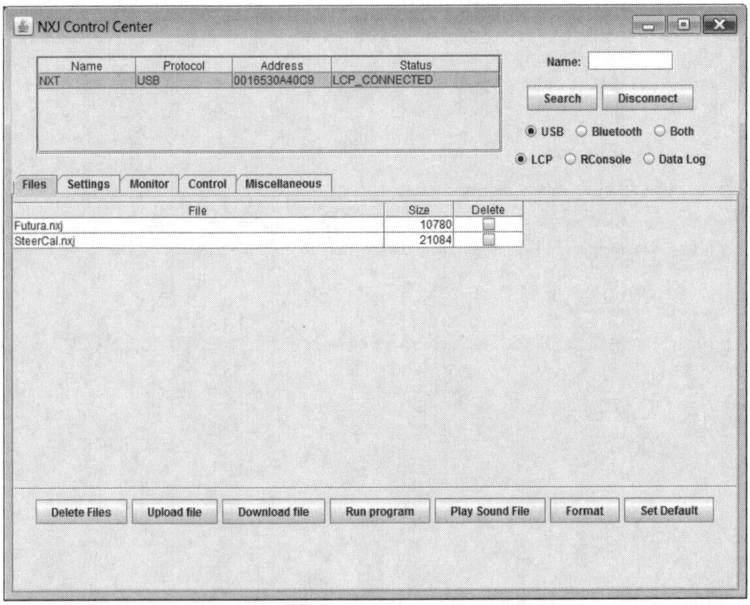

Figure 12-2: Browsing files on the NXT

There are also additional buttons. You can start a program on the NXT brick. Select it from the list by highlighting it (you do not have to place a checkmark next to it), then click Run program. You can also play a sound file through the brick, such as a wav file. The format button erases the entire file system (this does not delete the firmware).

Settings

The settings section allows you to change the NXT volume and the sleep delay, which determines the interval of inactivity to automatically power off the NXT brick (see Figure 12-3). If you want to play around with NXT Control Center without the brick accidently powering off, now would be a good time to raise this interval, to five minutes for example.

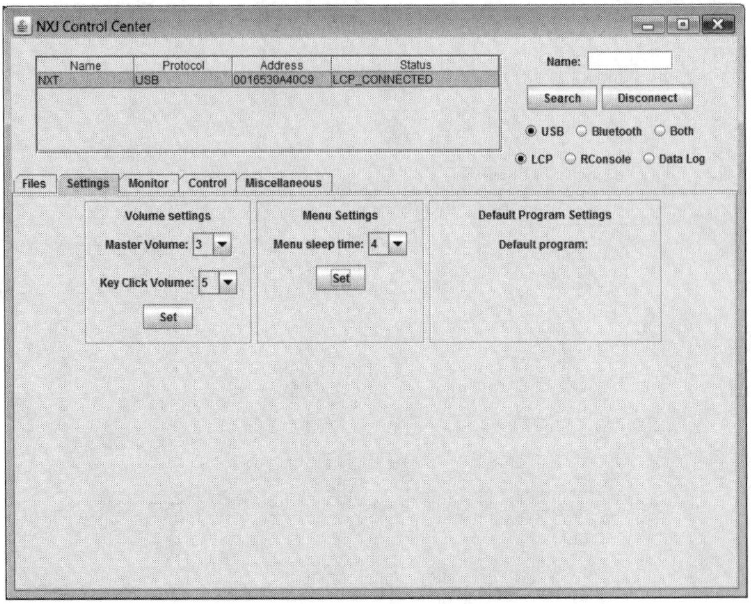

Figure 12-3: Changing the system settings

NOTE: The NXT brick will turn off after the default time if there is no activity. If you lose connection, your brick has probably turned off. You can also extend the automatic power-off time from the NXT menu. Enter the System menu, and then select Sleep Time. Every time you press the orange button it increases sleep time by a minute.

Sensor Monitor

The Monitor tab allows you to view real-time data from sensors connected to your NXT brick, as well as the battery. Try connecting a touch sensor to port 1. In the lower left corner, select S1, Touch Sensor, and Boolean (see Figure 12-4). Press and hold the switch on your touch-sensor and click update. You can see the raw value displayed at the top, while the scaled value (in this case, zero or one) is displayed on the bottom half of the panel.

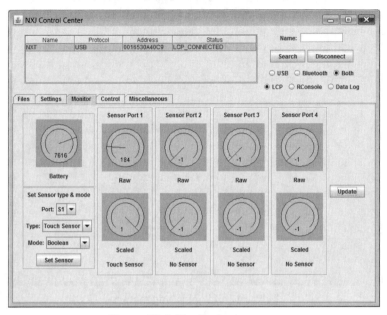

Figure 12-4: Monitoring sensors

Motor Control

A useful feature of NXJ Command Center is the ability to control motors without uploading an application (see Figure 12-5). This can help you when constructing a new model, allowing you to test different motors without writing a single line of code.

The motors will rotate for as long as you hold down one of the direction buttons (Forward or Backward). The Reverse checkboxes allow you to reverse the direction of one or more motors. The Limit field allows you to select the amount to rotate every time you hold down the button. For example, if you type 90 into the field, every time you hold down forward it will rotate up to 90 degrees and then stop.

You can also control a differential robot by selecting two motors, and then pressing Turn Left or Turn Right. This will cause the two motors to rotate in opposite directions of one another.

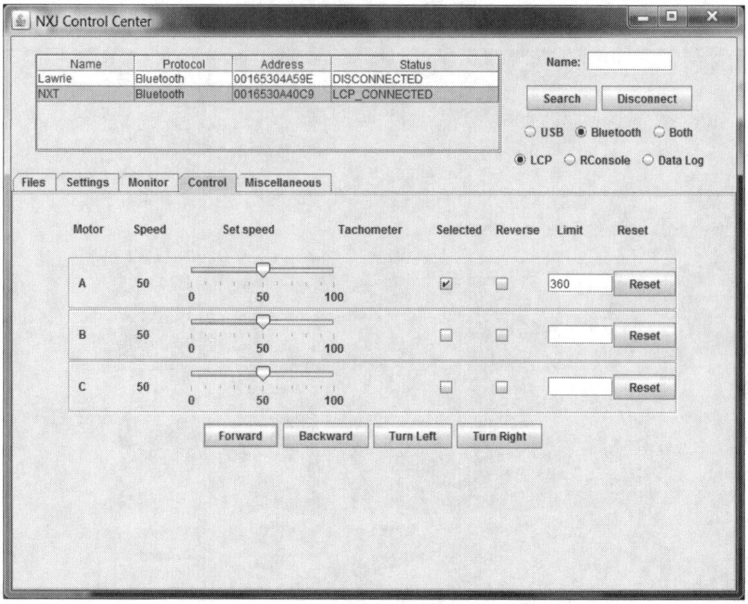

Figure 12-5: Controlling motors

Miscellaneous

The Miscellaneous tab has functions for playing a tone, changing the NXT brick name, and even sending test commands to I^2C sensors. This is useful if you are programming a new sensor and would like to see how different commands behave (these commands are usually documented in the product literature).

Data Logger

Have you ever looked at a project web page or Power Point presentation which contained a sample of data with a nice chart (see Figure 12-6)? They are actually easy to create by using a data logger, and this section will tell you how.

A data logger allows you to easily record and transmit data back to the PC. You could save your data to a file on your NXT brick using the File class, and then download it from the brick, but the data logger makes it a lot easier.

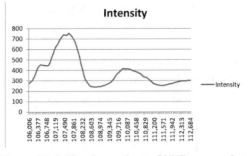

Figure 12-6: Viewing a chart of NXT sensor data

The Datalogger class contains methods to write float numbers to a data set. These methods all have the name writeLog() and you can insert one, two, three or four float parameters into this overloaded method. There is one other method called transmit() to send the data to the PC. Let's create a small program to record light readings and transmit them to the PC for display.

```java
import lejos.nxt.*;
import lejos.robotics.LightDetector;
import lejos.util.*;

public class LightLog {
  public static void main(String[] args) {
    Datalogger dl = new Datalogger();
    System.out.println("Press ENTER");
    Button.ENTER.waitForPressAndRelease();
    LightDetector ld = new
ColorSensor(SensorPort.S3);
    // LightDetector ld = new
LightSensor(SensorPort.S3);

    while(!Button.ENTER.isPressed()) {
      float val = ld.getNormalizedLightValue();
      int t = (int)System.currentTimeMillis();
      dl.writeLog(t, val);
      try {
        Thread.sleep(50);
      } catch (InterruptedException e) {
        e.printStackTrace();
      }
    }
    dl.transmit();
  }
}
```

This program will record light levels and the time each reading was taken.

1. To collect the data, plug a light or color sensor into port 3. NXT 1.0 owners will need to uncomment the sensor constructor line above and remove the ColorSensor reference.

2. Run the program and hit enter when you are ready to begin collecting data.

3. Aim the light sensor at a wall and slowly turn around so it takes about five seconds to complete one full turn.

4. When you have rotated 360 degrees, press the enter key.

5. The LCD display on the NXT brick will show a menu asking to transmit with USB or Bluetooth. Select whichever applies to you. It will now wait for a connection from the PC.

6. Now you need to connect to NXJ Control Center. Run NXJ Control Center and select the Data Log radio button.

7. Click Search and then select your brick from the list. Click Connect. The status will change to Datalog Connected (see Figure 12-7).

8. In the Columns field at the bottom of the window, enter 2, because we collected a set containing two data types. Now click Download.

9. Within moments you will see all the data on the screen (see Figure 12-8). Select all the data using your mouse, or you can hold down shift and use your arrow keys. Hit Ctrl-C to copy the data to the Windows clip-board.

10. Now open Excel or some other spreadsheet program. In the first row of the second column, type the word "Intensity".

11. Select the second row of the first column and hit Ctrl-V to paste the data into the spreadsheet. The data will now appear in the first two columns (see Figure 12-9).

12. Highlight the entire first two columns, then select the Insert menu and click on Line. It will instantly create a nice line chart as shown in Figure 12-10.

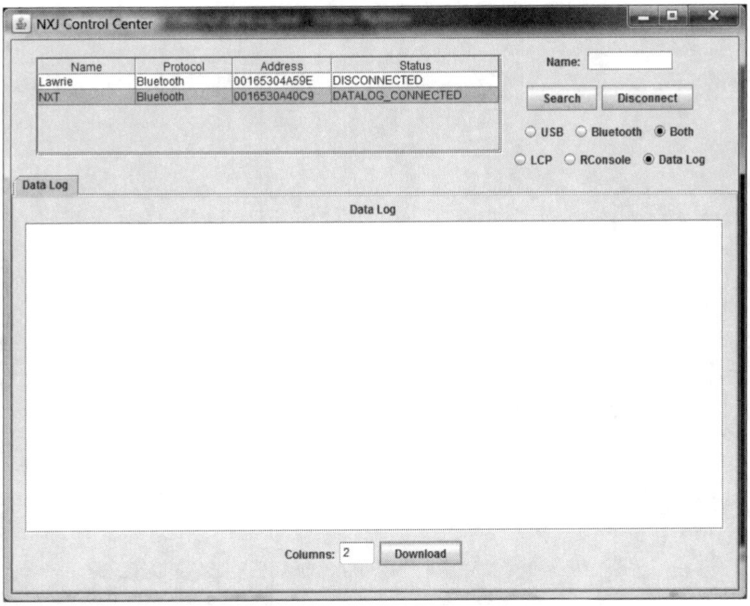

Figure 12-7: Creating a data log connection

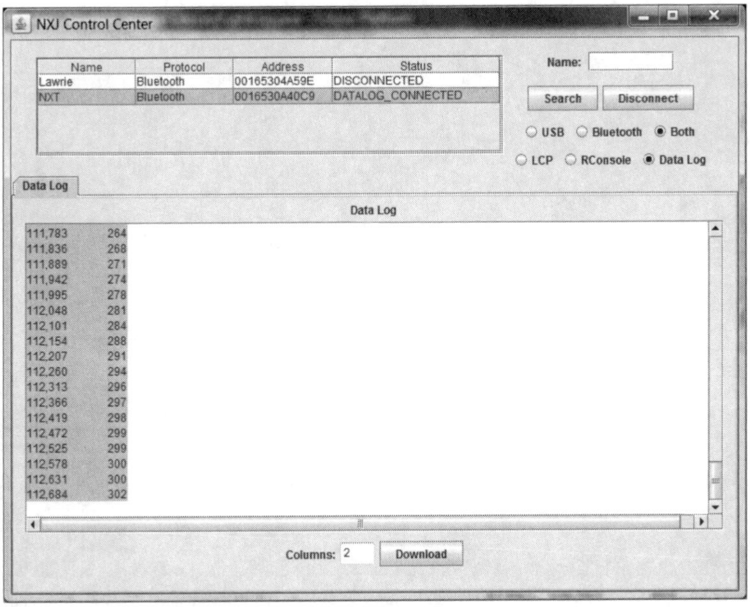

Figure 12-8: Viewing data in the Data Log viewer

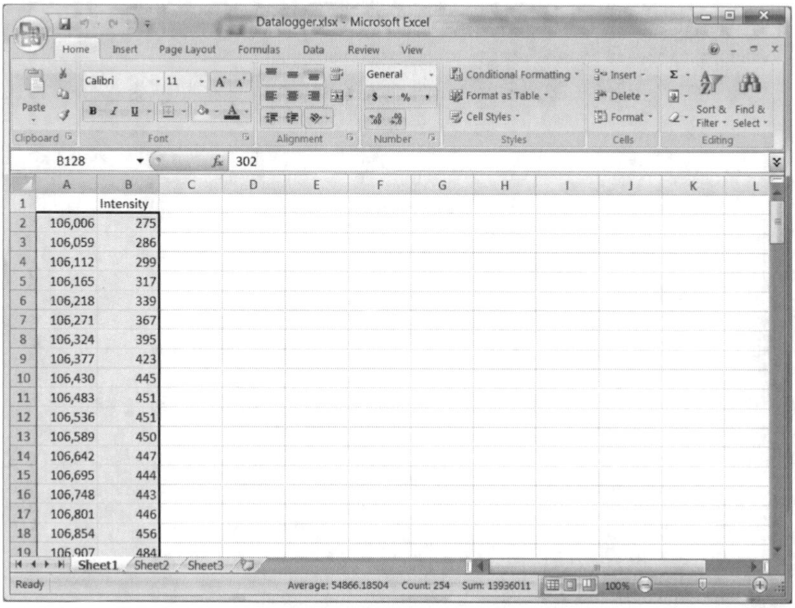

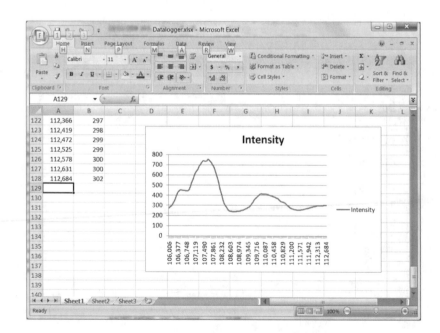

Figure 12-9: Pasting the data into a spreadsheet

Figure 12-10: Analyzing data in a chart

That's all there is to it! Now you can analyze the data. The first large bell-curve on my chart represents an office window, and the smaller bell-curve is the door, which was emitting slightly less light into the room than the window.

Remote Debugging

When programming, you sometimes want to output variables at certain parts of your code while it is running. Normally you can output the values to the NXT screen, but this screen is limited to eight lines of text. Also, if the program crashes, you will lose the text output before you can see what the values were.

The solution to this problem is to output text real-time to your PC. To do this, you need to add two lines to your code using the RConsole class. One line opens console output and the other closes the console. Let's examine some code to output text to the console.

```java
import lejos.nxt.comm.RConsole;

public class ConsoleOutput {

  public static void main(String[] args)
  throws Exception {
    RConsole.openAny(0);

    for(int i=0;i<100;i++) {
      System.out.println("Count " + i);
      RConsole.println("Counting " + i);
    }
    Thread.sleep(500);
    RConsole.close();
  }
}
```

This program counts from 0 to 99 and outputs the text to the LDC screen and the remote console.

1. Start ConsoleOutput on the NXT. It will display the message "Remote Console..."

2. Run NXJ Command Center on your PC and select the RConsole radio button.

3. Connect to the NXT brick as described above. Once connected, the program continues.

4. You will see a live screen capture of your LCD display on the bottom of the window, and text output to the console (see Figure 12-11).

This is a nice tool for viewing your LCD screen and debugging code. That's all there is to NXJ Command Center. If you are interested, feel free to check out the other utilities in the bin directory.

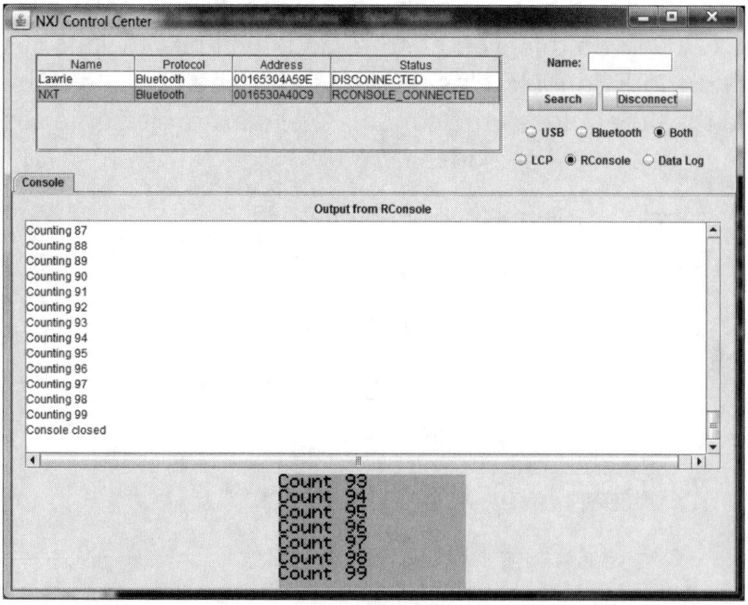

Figure 12-11: Viewing output from the NXT on your PC

CHAPTER 13

Maps

TOPICS IN THIS CHAPTER

- Maps
- Editing a map
- Controlling a robot using a map
- Whole home mapping

A map is a symbolic representation of an environment. It shows the relationship between objects on the map. The goal of this chapter is to create an accurate map of a test area, or perhaps even a large portion of your home. Once this map is complete, you can use the map to help guide your robot to specific locations. But first, let's look at how leJOS NXJ represents a map and the flexibility it provides in navigation projects.

Maps

As we learned earlier in the book, maps are an important tool for navigation (see Figure 13-1). They are a record of previous exploration, and can help you get to where you want to go. Maps are an important component of human navigation, such as with ships in the Navy, and they are important for navigational robots.

Figure 13-1: Using map data to navigate

Why create a map of the environment for your robot? There are two very good reasons, both of which are well worth the effort on their own. The first is that it is difficult to tell your robot where you want it to go. You can't just point with your finger to a spot on the floor and command it to go there. Remember how hard it was in earlier chapters to tell the robot where you wanted it to go using only coordinates? To get it to move to a specific point required taking out a ruler and measuring centimeters along the x and y coordinates. By viewing the map on a computer screen, you can very easily command the robot by clicking on different locations.

The second reason for making a map is that your robot can store the map in its memory and use the map geometry to plan a series of moves to a new location. The next chapter will demonstrate this concept in more detail. Even later, in chapter 26, your robot will use a map to find its location within an environment using a technique called Monte Carlo Localization. But for this chapter, we will deal with the first concept of telling a robot where to go on a map.

Autonomous navigation is often more complex than it first appears. It is one thing to program a robot to figure out how to physically get from one location to another, but sometimes the robot needs to navigate only where it is socially acceptable. For instance, the shortest route for a vehicle might be to cut through private property, but it is not socially acceptable (or legal) to do so. Maps can help to determine socially acceptable routes of navigation, not just physically possible ones.

The map classes in the leJOS API are mainly indoor specific. Outdoors, you'll find that detecting objects becomes difficult because terrain is uneven. It is a challenge for sensors to distinguish an impassable object from one that the robot can navigate over. Indoors is much simpler because the floor is a flat plane and all the walls are perpendicular to the floor, providing clear and unambiguous targets for the ultrasonic sensor to detect.

There are several ways a cartographer (ether amateur or professional) can choose to map an environment. We could use an occupancy grid map, which consists of a regular grid of squares (see Figure 13-2). But a simpler way to create a map is with a set of lines that represent walls. Most rooms have straight-line walls, so the leJOS API does not allow curved walls. You can still use diagonal lines for walls and objects in the environment, however.

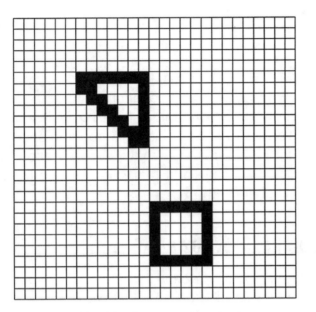

Figure 13-2: Viewing an occupancy grip map

The mapping API is located in lejos.robotics.mapping. Data is stored in a class called LineMap, and it simply consists of an array of lines. Each line consists of two coordinates—x1, y1, x2, y2. But how do we generate map data?

Editing a Map

The leJOS developers wanted to give users the ability to design a map using a dedicated map editor. Rather than including a proprietary map editor with leJOS, we decided to support standard file types for representing map geometry. That way, you can use the latest available map editing tools to create, save and edit your own maps.

It can take a long time to measure a room and enter it into an editor. With a good editor, lines and polygons can be moved, rotated, and named. For example, the living room might have furniture in it, such as a couch. If you rearrange your living room, it should be easy to load the map file into an editor and rearrange the polygons.

As of this writing, leJOS NXJ supports a few file types. The most important file format is called SVG, which stands for *Scalable Vector Graphics*. This is a popular open source standard for creating line geometry such as maps. This file format uses XML code and there are many editors available. Let's get started.

Creating a Test Area

Before you can begin creating a map, you need to decide on an area to map. We'll start small, because you will have to measure each wall of your test area. Robots are in their infancy still, so we want a simple, uncluttered area where the robot has little chance of running into dangerous obstacles. Think of it as a playpen for robots.

You can use household objects to close off a suitable area in one of your rooms, or use an existing walled area if it is completely enclosed by straight walls and is free from irregular furniture and other objects. For this chapter, I used the rear office wall and set up wooden partitions and filing cabinets to enclose a test area (see Figure 13-3).

I also tried to make sure the area was not symmetrical because I wanted to use the same map file and test area for Monte Carlo Localization later in this book. Once you have your test area ready, find a measuring tape and begin measuring the dimensions, preferably in centimeters. Record the measurements on a simple diagram. Nothing too fancy is required (see Figure 13-4).

Figure 13-3: Creating an enclosed test area

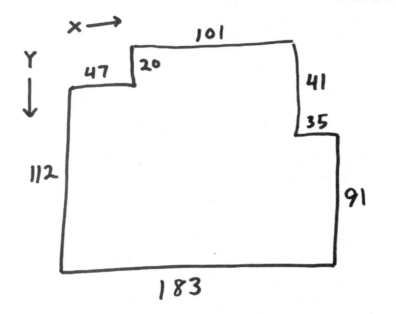

Figure 13-4: Recording wall measurements

Now that we have all the measurements, we can create our map.

Creating a Map

In this section we will create an SVG map using an editor. You could use any editor capable of using SVG files, such as Adobe Illustrator. However, for simplicity sake, we will use a dedicated web-based application hosted by Google called SVG-Edit. Because it is web-based, there is no installation required and it works on all platforms. You can find the editor at the following address.

http://code.google.com/p/svg-edit/

There are some excellent training videos available on the site, but they are largely overkill for our purposes. We just want to draw straight lines. To get started, go to the website and select the latest version (the screenshots in this chapter are from SVG-edit 2.6 running in a Firefox browser).

1. Click on the file menu, which looks something like a pencil drawing a flower (see Figure 13-5). Select Document Properties.

2. In the Document Properties window, enter a name for the map (see Figure 13-6). Enter the canvas dimensions to fit the size of your map area. The coordinate units in SVG are arbitrary, but we will consider them as centimeters. Uncheck the Snapping on/off option and click OK.

3. Select the line drawing tool and roughly draw one of the lines from your sketch. You do not need the exact line measurement when you sketch the line in because we will tweak the line dimensions. You can manually enter the proper coordinate values at the top of the screen in the x1, y1, x2, y2 fields (see Figure 13-7).

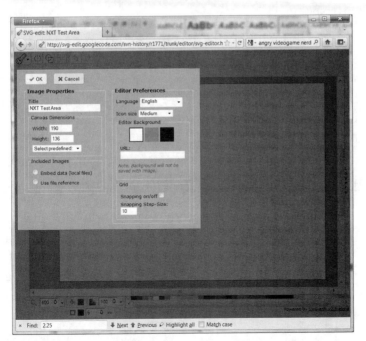

Figure 13-5: The file menu

Figure 13-6: Changing the document properties

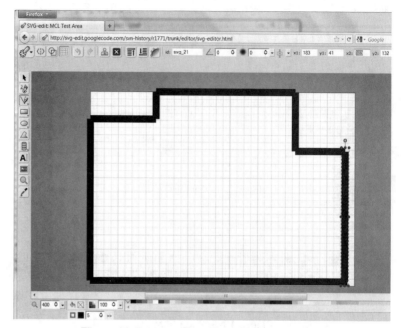

Figure 13-7: Manually editing line coordinates

4. Continue drawing all the lines of your map. When you are done, click the menu icon and select Save Image. In Firefox, a new window appears showing the map. You can either right click the map image and select Save Page As, or select File > Save Page As... Remember where you saved the file as we will copy it to our project folder later in this chapter.

Controlling a Robot using a Map

In this section we will create a very simple but powerful NXT application that will display real-time map data on a PC. You can make your robot travel, rotate and go to different coordinates by clicking on a point on your map to make the robot drive to that location. At all times you will see the present location of the robot and the path it has driven.

NXT Program

The NXT program is very simple. It creates the usual navigational classes, such as a pilot and a navigator. Once this is done, it waits for the PC to connect to it using the class NXTNavigationModel.

The NXTNavigationModel class allows you to send navigation commands to your robot, but also to synchronize data (such as pose) with the PC. The code below shows how to add a Pilot and Navigator to the NXTNavigationModel. Once these are added, the PC map application will visually update all the movements of your robot on the display.

```
import lejos.nxt.Motor;
import lejos.robotics.mapping.
NXTNavigationModel;
import lejos.robotics.navigation.*;

public class NXTSlave {
  public static void main(String[] args)
throws Exception {
    DifferentialPilot robot = new Differential
Pilot(5.6,16.4,Motor.B,Motor.C);
    robot.setAcceleration(500);
    Navigator navigator = new Navigator(robot);
    Pose start = new Pose(17, 97, 0);
    navigator.getPoseProvider().setPose(start);
    NXTNavigationModel model = new
NXTNavigationModel();
    model.addPilot(robot);
    model.addNavigator(navigator);

model.addPoseProvider(navigator.
getPoseProvider());
  }
}
```

This code uses a differential robot, such as the Carpet Rover (see Figure 3-3). Make sure to modify the code above with the proper tire size and track width for your robot. You will also need to decide the starting coordinates of your robot within your test area. You might want to place a coin at this location so you don't forget it for subsequent trials. Measure the coordinates of this location and then change the following line of code to include the x, y coordinates and heading.

```
Pose start = new Pose(17, 97, 0);
```

Upload and run the code. Place your robot in the test area at the starting pose you specified in the code. Now we need to run an application on the PC.

PC Program

The leJOS developers have included a comprehensive application for displaying map data and controlling navigators on a PC. The application, Map Command, is located in the bin directory of your leJOS installation. The file is nxjmapcommand.bat. Double click this file to run it and you will see a screen with a grid. Once you load in your map, you will see it displayed on screen. Use the zoom feature to adjust the view so your map takes up most of the screen (see Figure 13-8).

> **NOTE:** Map Command might not appear exactly as shown in the screenshots below due to version revisions.

Enter the name of your NXT brick in the connect field and press Connect. When the PC and NXT have established a Bluetooth connection you can begin making moves. Right click any part of the map area and select Go To. The robot will drive to that point. When it does, the GUI displays a trail where it has been (see Figure 13-9). You can also manually control the robot's travel distance and angle.

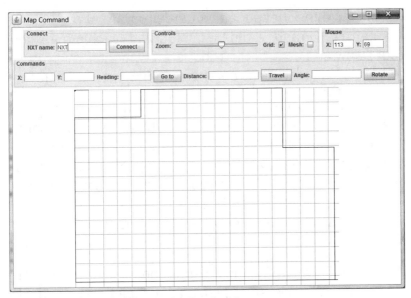

Figure 13-8: Adjusting the map zoom

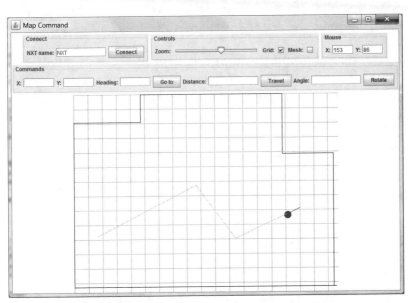

Figure 13-9: Traveling to coordinates on the map

Whole Home Mapping

To maximize the functionality of the program above, a larger multi-room map would be useful. With such a map, you could view and control the robot even when it is away from your view—so long as it remains within Bluetooth distance.

The real chore is making a large map of your home. To speed this process up, you can retrieve the blueprints of your home (see Figure 13-10). They should be available from the local government in your city. If you have a more sophisticated editing tool, you can scan in your blueprints, scale them to the proper size, and then trace the outer edges of rooms to create lines. When you are done, delete the blueprint layer from your image and export the line drawing as an SVG file.

One problem with mapping an entire home is that some objects are semi-permanent, such as a chair. They are likely to move around. Also, some objects are at angles for the scanner on your small robot. For most tables and chairs, only the legs are visible to an ultrasonic sensor, so your map will have to represent them as four small squares.

One bonus of going through this exercise is that it will motivate you to clean your house room by room because you don't want junk on the floor to confuse the robot. Have fun!

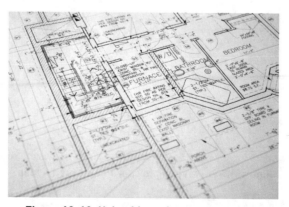

Figure 13-10: Using blueprints to map rooms

CHAPTER 14

Pathfinding

TOPICS IN THIS CHAPTER

- Nodes
- Search Algorithms
- Navigation Meshes
- Adding obstacles to your map

Pathfinding is an interesting problem that has been largely conquered in the field of computing. The goal is to find the shortest path from one point to another, regardless of the obstacles that stand between these two points (see Figure 14-1). There are discrete concepts of pathfinding, including search algorithms, nodes, and navigational meshes. Let's explore these concepts.

Pathfinding

Humans have the built-in ability to quickly find the shortest path from one point to another. Take a look at Figure 14-1. Can you trace the shortest path to the goal? It probably took you no thought at all. Surprisingly, humans have a hard time describing *how* they did it—that is, the thought process behind how they arrived at the solution. It is especially difficult for a programmer to describe to a computer (in the form of a program) how to find the shortest path.

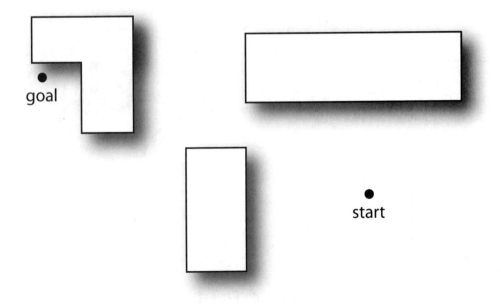

Figure 14-1: Traveling from one point to another

Before we can instruct a computer how to perform path-finding, we need to develop an abstract representation of the environment that a computer program can understand. One component of this model consists of a map representation, as we explored in the previous chapter. The other component of pathfinding is a node—or rather, a set of nodes.

Nodes

Imagine a dense, impenetrable forest with lots of paths snaking throughout it. These paths occasionally link-up with other paths (see Figure 14-2). Wherever two or more paths link together, we have the option of going down one of two or more paths. We must make a decision which path to take. The places where these paths cross are called nodes, and they are a central concept of pathfinding.

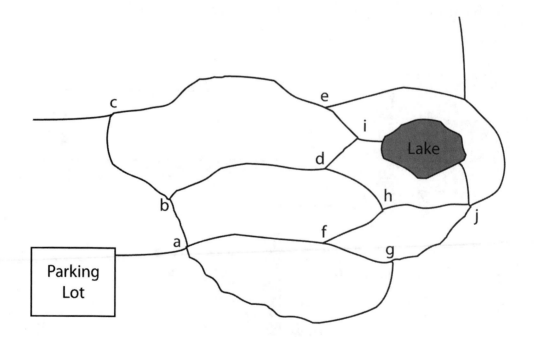

Darkplace Forest

Figure 14-2: Exploring a dense network of forest paths

A node is a variation of a coordinate. In a two dimensional space, each node has an x and y coordinate. So how do nodes differ from coordinates? Each node is connected to at least one other node (see Figure 14-3). With the node model, there is no option to break away from one path onto another path. You can only change direction when you reach a new node.

In Figure 14-2, if you want to plot a route from the parking lot to the lake, there are a multitude of possible routes. The shortest path is A to B to D to I. Our goal is to have the computer calculate this type of path for our robot.

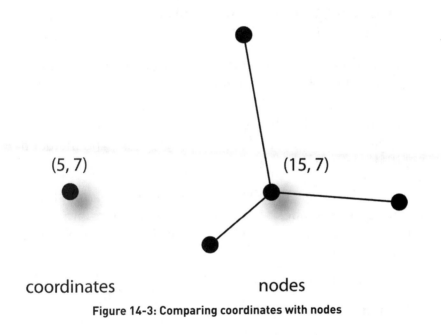

(5, 7)　　　　　　　　　(15, 7)

coordinates　　　　　　　　nodes

Figure 14-3: Comparing coordinates with nodes

Navigation Mesh

The nodes in the forest example above were all predetermined for you. You did not design the nodes because the forest paths were already long established before you arrived at the scene. The crisscrossing roads in a city are also predetermined for you, with every intersection representing a node where your vehicle can take two or more different branches.

But what about a relatively open environment where there are no predetermined paths? Most indoor robotics problems consist of a relatively wide open area with some sporadic obstacles in the environment (see Figure 14-1). Where are the nodes and adjoining paths? As you can see, no predefined nodes exist in this type of environment as they do with roads or forest paths.

This problem leads us into the interesting problem of navigation meshes. To automate pathfinding, we ideally want to feed the robot a map and then let it figure out how to get from one point to another. So without any nodes, it needs to generate a set of nodes on its own. This complete set of nodes, which covers the entire area of a map, is called a *navigation mesh*.

Navigation meshes are used extensively in video games—normally first person shooters like Call of Duty and real-time strategy games like Starcraft—to help characters move from one location to another. Usually the navigation meshes are pre-calculated by the designers and included with the map data for a particular level.

Similarly, with a typical indoor environment, we must generate a set of nodes to overlay the map. There are many different ways you can generate a navigation mesh. A few common ones include:

- regular grid pattern
- randomly generated
- heuristically generated

There are downsides to some types of navigation meshes. For example, the grid mesh produces paths that are all perpendicular to one another without diagonal paths, which results in, well, very robotic movement by your robot. Figure 14-4a shows the shortest path a robot might take using a grid mesh. However, there are shorter paths that you and I would take if we weren't locked onto a grid mesh, such as that shown in Figure 14-4b. By generating an optimal navigation mesh for a particular environment, we can find a shorter overall path.

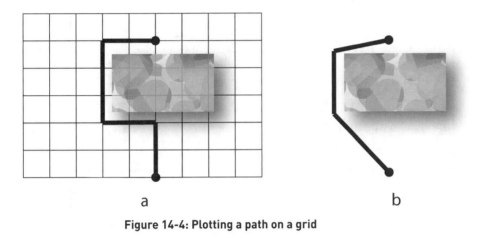

a   b

Figure 14-4: Plotting a path on a grid

As you can see above, there is a lot of empty white space within navigation meshes. Technically a more complete navigation mesh would include lots of interconnected nodes all across the map. For perfect coverage, you could include one node spaced at the atomic level. However, this is not practical as it would use too much memory and produce very long searches.

The best we can do is populate the space with a diverse selection of nodes, which is merely an approximation of the number of places the robot could navigate to on the map. As you can see, there are limitations with the node-search paradigm.

One other factor of consideration when generating a navigation mesh is node pruning. There are often illegal paths between nodes that are invalid because they would cause the robot to collide with walls or other objects (see Figure 14-5a). If you are plotting a path for a boat, you probably want to stay away from islands, or perhaps stay in deep water channels if the topography below the surface is known. Therefore, you might want the node generation routine to create a navigation mesh that takes these factors into account, rather than relying on a grid or random node set.

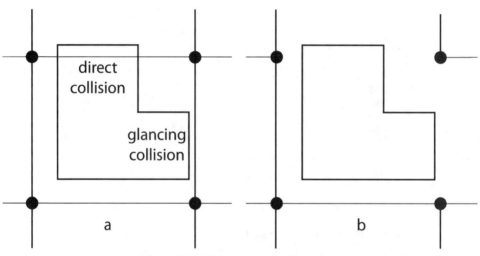

Figure 14-5: Pruning connections from a set of nodes

By checking the connections between nodes for collisions with the map geometry, we can tell which connections should be pruned. Then we merely remove these connections between the nodes (see Figure 14-5b). The leJOS node generation algorithm uses a pruning algorithm to keep a wide berth between your robot and other objects.

Search Algorithms

The final concept we will cover in the pathfinding topic is node searching. A node search uses an algorithm to search though a set of nodes to find the optimal path from a starting node to the goal node.

Modern search algorithms started with Dutch mathematician Edsger Djikstra in 1956. He wanted an algorithm to find the shortest path through a maze. The algorithm he came up with—today called Djikstra's algorithm—is proven to find the shortest path between any two nodes.

Because Djikstra's algorithm finds the most optimal path between nodes, you might think that there is no point in doing any further research into node searches. After all, nothing can beat the path it generates. However, in the 1960s, his algorithms saw practical use with early computer networks, specifically in routers. Variations of his algorithm are used by network routing protocols to make the Internet faster by finding the shortest route for data packets between nodes (see Figure 14-6). Obviously engineers want fast data transmission, so they set to work on improving the speed of these search algorithms.

Since then, a whole family of search algorithms have been produced, the most famous being A* (pronounced A star). A* can improve search speeds by a factor of 20 times or more over Djikstra's original algorithm.

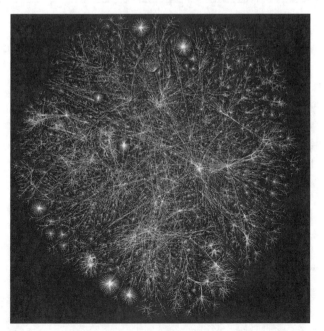

Figure 14-6: Data packets travelling between Internet routers

Node search algorithms are mainly used in video games for pathfinding, consumer GPS devices to plot routes, web-based mapping applications such as Google maps and of course, mobile robots. The Mars Pathfinder mission, launched in 1996, used D* pathfinding. Later, a variation on this, called D* lite, superseded D* to become the most popular search algorithm for use in robotics.

The bottom line is that all of these search algorithms produce the same results. It is a matter of how fast they produce results, and how many resources they use in terms of memory and computing horsepower. This can be a factor for devices with relatively slow processors and limited memory, such as the NXT brick.

Let's examine a pathfinding example on your NXT brick. First we will need to alter the environment and map data from chapter 13.

Adding Obstacles to a Map

Pathfinding is not very interesting unless there are obstacles between the starting node and the goal node. Add one or more obstacles to the test area from chapter 13. To keep things simple, I added a single box in the middle of my test area (see Figure 14-7). Measure the dimensions of the box and the relative location of the box compared to two perpendicular walls. Now we will need to add this obstacle to our existing map data.

Let's return to SVG Edit from the previous chapter and modify the map we created.

1. Launch the SVG-Edit application at:

   http://code.google.com/p/svg-edit/

2. Click on the menu icon and select Open Image (see Figure 14-8). Do not select Import SVG as this will add the SVG geometry to the current blank document.

Figure 14-7: Adding an obstacle in the test area

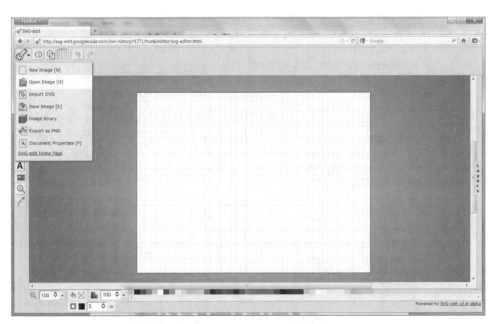

Figure 14-8: Opening the map data in SVG Edit

3. Using your measurements, add the new obstacle(s) to your map (see Figure 14-9).

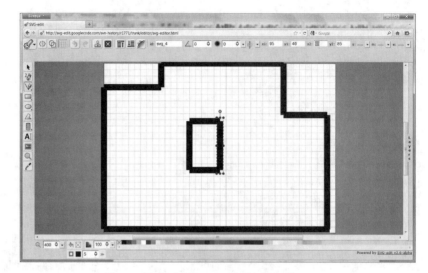

Figure 14-9: Adding an obstacle to the map

4. Alternately, you could create a larger floor plan to give the pathfinding algorithms a real challenge (see Figure 14-10).

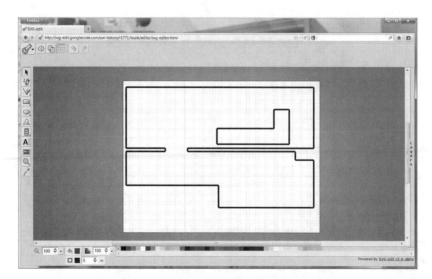

Figure 14-10: Creating a larger floor plan

When completed, save your map file to a folder or your desktop. We will use this map later in this chapter.

Programming Pathfinding

The application in this section is a culmination of all the ideas we explored in the navigation chapters. It uses not only a Pilot, Navigator, PoseProvider (hidden) and map data, but also a PathPlanner. In effect, all the key players are onboard the same boat (see Figure 14-11).

Figure 14-11 Working together on a navigation problem

The following application connects to the NXT and instructs the robot to follow a calculated path. It works best with a large map, such as a whole floor with doorways rather than a single room. The initial pose and the target are set using popup menus.

There is no new code for this project. We simply use Map Command on the PC side and the NXTSlave code from the previous chapter. However, we will select different Map Command options this time.

1. Run nxjmapcommand.bat and load in your SVG file from the previous section. Resize the zoom bar until the map fits your screen (see Figure 14-12). Make sure both grid and mesh are selected.

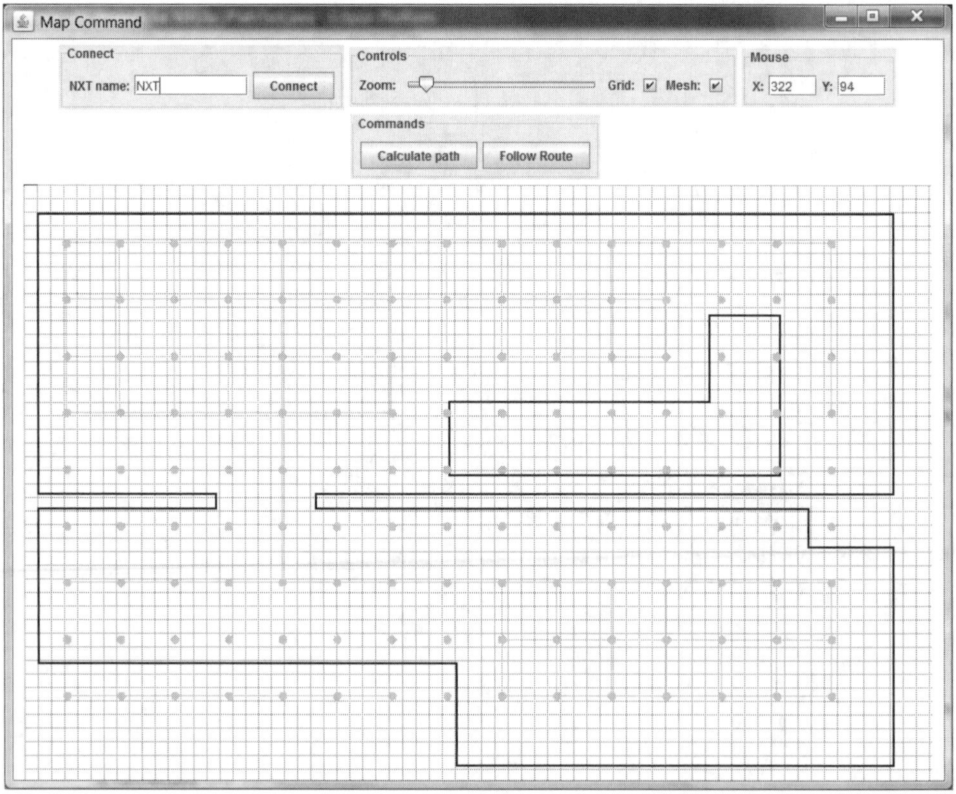

Figure 14-12: Displaying the map data

2. Set down your NXT robot at a location within your test area and note the coordinates. Now run the NXTSlave program.

3. On Map Command, enter the name of your NXT brick and click Connect.

4. Now move the pointer to the starting location of your robot on the map. When you see the correct coordinates in the upper right corner, click the grid and select Set Pose. A new node will appear at the location connected to the nearest nodes (see Figure 14-13).

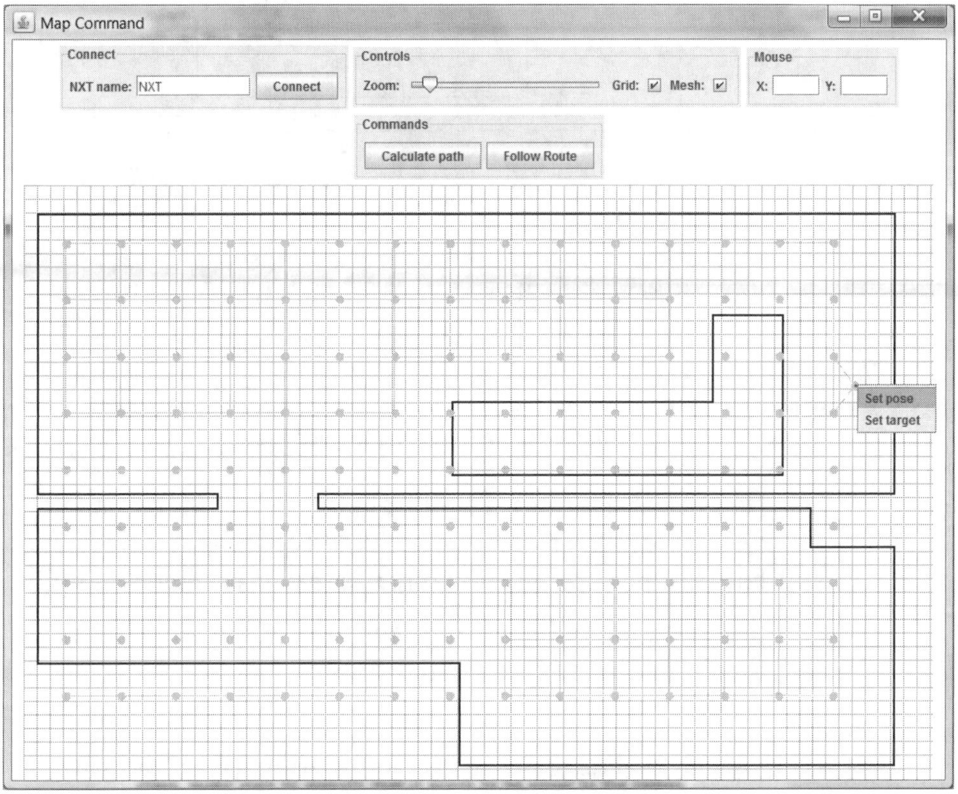

Figure 14-13: Selecting the starting node

5. Now move the pointer to a destination on the map, click the mouse and select Set Target. You should now see two new nodes on the map (see Figure 14-14).

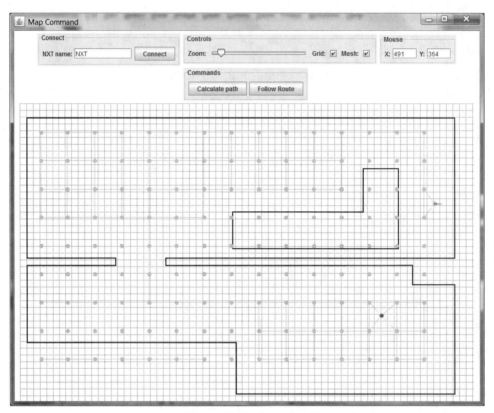

Figure 14-14: Selecting the destination

6. Now click Calculate Path. The algorithm will quickly find the shortest path between the robot and destination nodes. A line will show the path the robot will take (see Figure 14-15).

7. Finally, click Follow Route and your robot will begin moving to the target destination.

That's a simple demonstration of the pathfinding algorithm. You can also access the pathfinding classes directly and program your own custom applications. You can examine these classes in the lejos.robotics.pathfinding package.

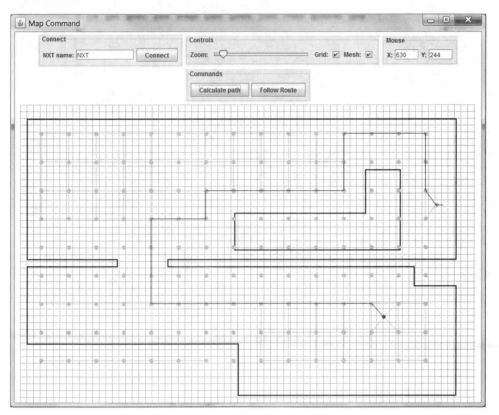

Figure 14-15: Generating a path

CHAPTER 15

Object Detection

TOPICS IN THIS CHAPTER

- Object Detection
- Feature Detectors
- Feature Listeners

The leJOS developers wanted a standard set of classes to detect objects and obstacles. We came up with the package lejos.robotics.objectdetection in order to generalize the task of object detection. We want to allow our navigation classes to deal with objects that might be detected in the environment. But since there are multiple sensors and methods for detecting objects, it is important for our API to be able to deal with not only standard object detection sensors (such as ultrasonic sensor) but also custom systems users might create on their own. With a generic object detection API, a user can hack together as many object detection strategies they like and use them in our standard navigation classes.

Object Detection

Object detection allows a robot to detect objects in its path and take some action, such as avoiding the object. We chose the word feature to describe an object rather than obstacle because sometimes the robot is seeking an object out, rather than avoiding it (such as a soccer ball or robot combatant). The word object is not used because it was a potentially confusing class name with object oriented Java programming.

Feature Detectors

A FeatureDetector reports on objects it detects using a sensor, such as a touch sensor or ultrasonic sensor. It is the main interface in the object detection package from which data originates. There are many benefits of using a FeatureDetector:

- Automatic scanning and reporting of data
- A listener interface to segregate the action-response code
- Allows a single block of code to respond to data from multiple sensors

Two implementations of FeatureDetector in our API are:

- RangeFeatureDetector - uses RangeFinder classes, such as the LEGO ultrasonic sensor.
- TouchFeatureDetector - uses Touch classes, such as the LEGO touch sensor.

The RangeFeatureDetector allows you to choose some parameters for the sensor, such as the maximum range you want it to report findings for, and the time between performing scans. You can construct a simple RangeFeatureDetector as follows:

```
int MAX_DISTANCE = 50; // In centimeters
int PERIOD = 500; // In milliseconds
UltrasonicSensor us = new
UltrasonicSensor(SensorPort.S4);
FeatureDetector fd = new
RangeFeatureDetector(us, MAX_DISTANCE, PERIOD);
```

Once you have a FeatureDetector instantiated, such as fd above, you can perform a scan on it to retrieve data:

```
Feature result = fd.scan();
if(result != null)
  System.out.println("Range: " + result.
getRangeReading().getRange());
```

NOTE: Make sure to check for a null object before trying to read data from the returned Feature object, otherwise your code will throw a null pointer exception.

The main benefit of a FeatureDeteector is the ability to automatically notify other classes when an object is detected via a listener interface. The next section discusses this in more detail.

Feature Listeners

Once you have a FeatureDetector instantiated, you can add a FeatureListener to it. The FeatureListener code is notified when an object is detected via FeatureListener.featureDetected(). By implementing this method, you can make your robot react to the detected object by performing some sort of action. The following code shows how to use a FeatureListener:

```
RangeFeatureDetector fd = new
RangeFeatureDetector(us, MAX_DETECT, 500);
fd.addListener(listener);
```

A more complete example of using a FeatureListener is included in the full code sample below.

Feature data

As you saw above, data is returned from FeatureDetectors in a Feature object. This is actually an interface which defines the basic requirements of a Feature object. The most basic class to implement the Feature interface is the RangeFeature class. This class is a data container, and the data is retrieved by these methods:

```
getRangeReading()
```

• Returns a RangeReading object

```
getRangeReadings()
```

• Returns a RangeReadings collection

```
getTimeStamp()
```

• Returns the system time (in ms) when this data was collected

Some scanners are capable of detecting multiple objects, such as the LEGO ultrasonic sensor. Other sensors are only capable of detecting a single object, such as a touch sensor. In either case, they are both capable of producing data to fulfill the first two methods listed above.

How? When the getRangeReading() method is called, it returns the closest object that was detected by a scanner, even if it is capable of returning more than one hit.

When the getRangeReadings() method is called, it returns all objects it detected. If the scanner is only capable of returning one hit, the RangeReadings object will contain only one RangeReading.

Handheld Range Meter

Let's create a simple but fun project to demonstrate the object detector and listener interface. The device is simply an ultrasonic sensor attached to the NXT brick (see Figure 15-1). Using a short cable, plug the sensor into port 4.

Figure 15-1: Building the simplest project in the book

The code below uses the RangeFeatureDetector constructor to limit the distance objects are reported. The code uses 150 cm, but you can change this as you wish. By default, it reports readings every 500 ms, but you can change this too.

```
import lejos.nxt.*;
import lejos.robotics.objectdetection.*;

public class ObjectDetect implements
FeatureListener {

  public static int MAX_DETECT = 150;

  public static void main(String[] args)
throws Exception {

    // Instructions:
    System.out.println("Autodetect ON");
    System.out.println("Max dist: " + MAX_DETECT);
    System.out.println("ENTER = do scan");
    System.out.println("RIGHT = on/off");
    System.out.println("ESCAPE = exit");
```

```java
   // Initialize the detection objects:
   ObjectDetect listener = new ObjectDetect();
   UltrasonicSensor us = new
UltrasonicSensor(SensorPort.S4);
   RangeFeatureDetector fd = new
RangeFeatureDetector(us, MAX_DETECT, 500);
   fd.addListener(listener);

   // Disable default button sound:
   Button.setKeyClickVolume(0);

   // Button inputs:
   while(!Button.ESCAPE.isPressed()) {

     // Perform a single scan:
     if(Button.ENTER.isPressed()) {
       Feature res = fd.scan();
       if(res == null) System.out.
println("Nothing detected");
       else {
         listener.featureDetected(res, fd);
       }
       Thread.sleep(500);
     }

     // Enable/disable detection using buttons:
     if(Button.RIGHT.isPressed()) {
       if(fd.isEnabled()) {
         Sound.beepSequence();
         System.out.println("Autodetect OFF");
       } else {
         Sound.beepSequenceUp();
         System.out.println("Autodetect ON");
       }
       fd.enableDetection(!fd.isEnabled());
       Thread.sleep(500);
     }
     Thread.yield();
   }
 }

 public void featureDetected(Feature feature,
FeatureDetector detector) {
   int range = (int)feature.
getRangeReading().getRange();
   Sound.playTone(1600 - (range * 10), 100);
   System.out.println("Range:" + range);
 }
}
```

Just for fun!

The handheld range meter is an interesting project for testing the performance of the ultrasonic sensor. Try pointing it at a wall and then changing the angle (see Figure 15-2). Notice that if you go over a certain angle it no longer returns a reading? That is because the sound is being deflected away from the sensor instead of reflecting back towards it. Also, try pointing it at a soft or furry object, like a dog. Dogs are almost "radar invisible" to the ultrasonic sensor, much like a stealth plane.

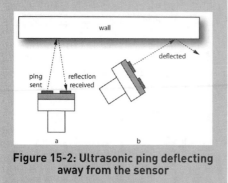

Figure 15-2: Ultrasonic ping deflecting away from the sensor

Once you run the program you can walk around your house like Spock with a tricorder, pointing it at different objects and watching the distance readout on the LCD. You can also disable the auto-detect listener by pressing the right arrow button. In this mode, simply press enter to take a single reading.

Rear Bumper Robot

It is also possible to combine several different FeatureDetectors into one FeatureDetector using the FusorDetector class. This is useful for robots that have a number of sensors located around the robot, such as several bumpers at different locations, plus several range sensors. By doing this, it allows one FeatureListener to respond to all of the sensors of the robot.

To test this out, we'll need to mount at least two sensors to a robot. This example uses the Carpet Rover with an ultrasonic sensor and a rear bumper. Start with the Carpet Rover equipped with an ultrasonic sensor (see Figure 4-6). Then continue with the instructions below:

1

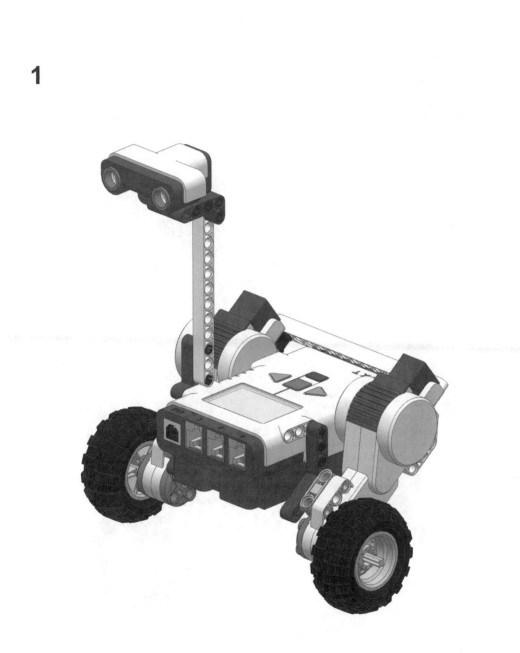

2

4x

2x

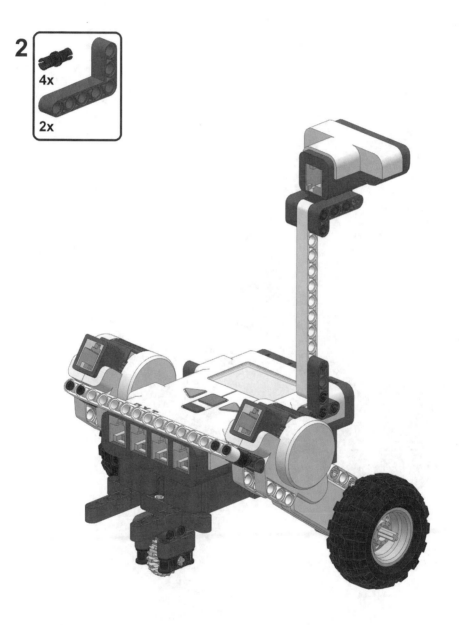

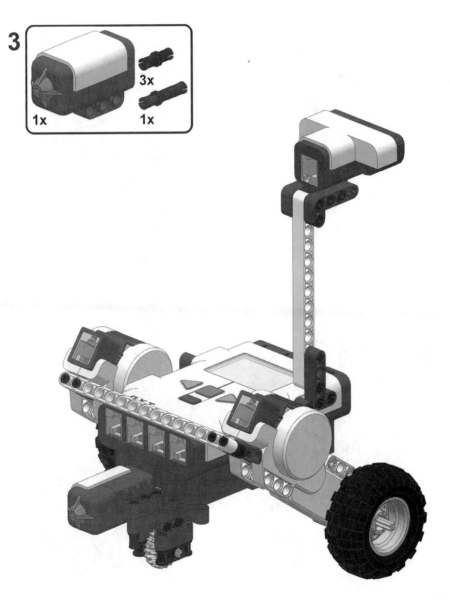

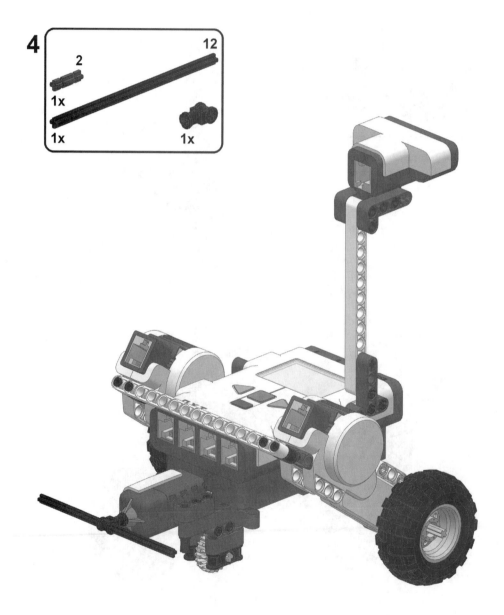

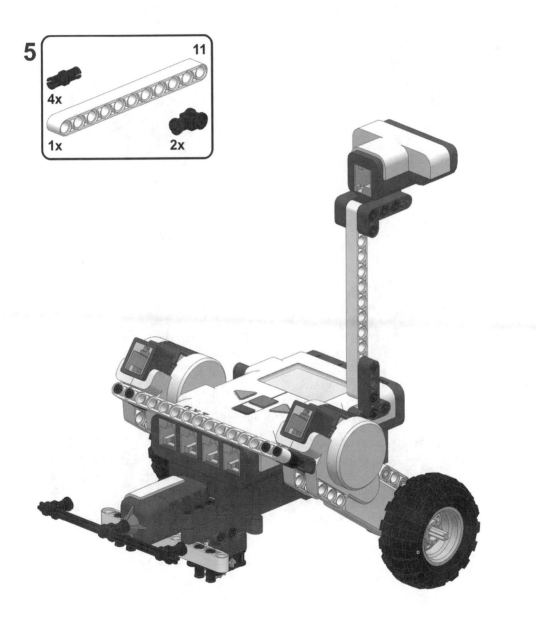

6

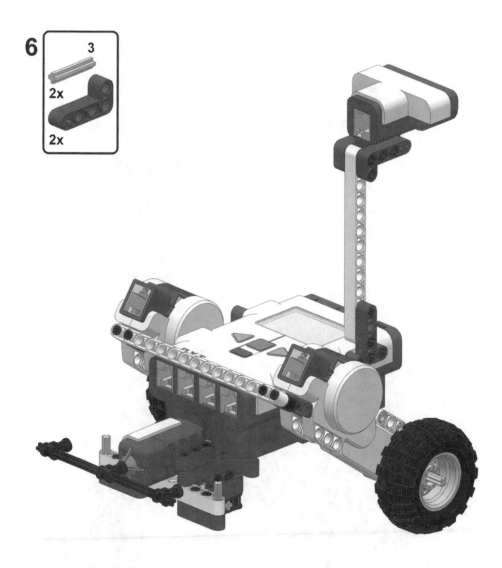

Now enter the following code and upload it to the robot.

```
import lejos.nxt.*;
import lejos.robotics.Touch;
import lejos.robotics.navigation.
DifferentialPilot;
import lejos.robotics.objectdetection.*;

public class BumperBot implements
FeatureListener {

  private static final int MAX_DETECT = 50;
  private static final int RANGE_READING_
DELAY = 500;
  private static final int TOUCH_X_OFFSET = 0;
  private static final int TOUCH_Y_OFFSET = -14;

  private DifferentialPilot robot;

  public BumperBot() {
    robot = new DifferentialPilot(4.32, 16.35,
Motor.B, Motor.C, false);
    robot.forward();
  }

  public static void main(String[] args )
throws Exception {
    UltrasonicSensor us = new
UltrasonicSensor(SensorPort.S4);
    FeatureDetector usdetector = new
RangeFeatureDetector(us, MAX_DETECT,RANGE_
READING_DELAY, 90);

    Touch ts = new TouchSensor(SensorPort.S1);
    FeatureDetector tsdetector = new
TouchFeatureDetector(ts, TOUCH_X_OFFSET,
TOUCH_Y_OFFSET);

    FusorDetector fusion = new
FusorDetector();
    fusion.addDetector(tsdetector);
    fusion.addDetector(usdetector);

    fusion.addListener(new BumperBot());

    Button.waitForAnyPress();
  }
```

```
  public void featureDetected(Feature feature,
FeatureDetector detector) {
    System.out.println("R:" + feature.
getRangeReading().getAngle());
    if(feature.getRangeReading().getAngle() >=
0) {
      detector.enableDetection(false);
      robot.rotate(90 * Math.random());
      detector.enableDetection(true);
      robot.backward();
    } else {
      robot.forward();
    }
  }
}
```

Plug the touch sensor into port 1 and the ultrasonic sensor into port 4. This code makes the robot drive forward until it detects an obstacle. Then it rotates to a new direction and reverses until the rear bumper detects an object.

Watch the bumper car carefully to see how long it survives before something goes wrong. Try to note what it is that goes wrong. Alter the code to make the robot more robust as it navigates. Or, maybe you need to physically modify the bumpers? The goal is to create consistency, repeatability, longevity and robustness in your robots.

CHAPTER 16

Kinematics

TOPICS IN THIS CHAPTER

- Robotic Arms
- Kinematics
- Calculating Inverse Kinematics

Robotic arms are commonly used in manufacturing. For example, the cars in your neighborhood were welded and partially assembled by industrial robotics. This chapter will explore the concepts needed to successfully build and program arms that can accurately navigate 3D space.

Robotic Arms

The human arm is a relatively simple piece of machinery compared to the human hand. The arm is like a big finger, with a shoulder joint that can rotate along two axes, and an elbow that can rotate along one axis. Like other parts of your body, the arm uses proprioception. In other words, you can sense the position of your arm even when you are not looking at it.

There are at least two joints in a robotic arm, the shoulder and the elbow. The shoulder is sometimes capable of rotation in two axes, which requires two dedicated motors. For example, the robot we will build in this book, LARA, has three axes of rotation (see Figure 16-1). As a result, robotic arms commonly require three motors—not including motors that control the end effector, such as a claw.

Often robotic arms do not mimic human arms, such as the popular manufacturing arm known as SCARA (see Figure 16-2). A SCARA arm is only capable of moving along two dimensions (x and y) and not up and down.

Just for fun!

Try paying attention to your own sense of proprioception. First, place a nonbreakable object on the table in front of you and hang your hand down by your side. Look at the object, close your eyes and then try to pick up the object. Chances are you were able to pick it up on the first try. Your body tells your brain the approximate angle of each joint in your arm, allowing you to maneuver your arm into position without looking. Even though the robot arm described in the next chapter is blind, it can accurately move to 3D coordinates just like your own arm.

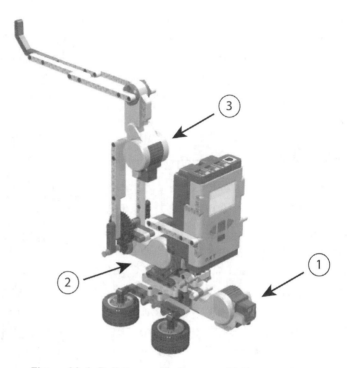

Figure 16-1: Building a robotic arm with three motors

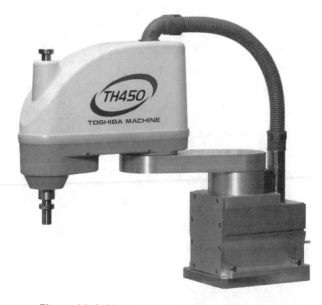

Figure 16-2: Manufacturing with a SCARA arm

Now that we know the constituent parts of an arm, we need to figure out how to allow our robot to move its end effector to a specific location, the same way you were able to move your hand to an object with your eyes closed. We'll use a mechanical engineering discipline known as kinematics.

Kinematics

Kinematics is a branch of mechanics that can describe the motion of bodies. Specifically, kinematics is good at describing rigid bodies that are linked together at rotating joints. We can use kinematics to calculate the position of a robot hand, or we can instruct the arm to move a hand to a specific location in 3D space. There are two sub-branches of kinematics of interest for robotics.

The first is *forward kinematics*, which calculates the position of a robot hand given the angles of the joints (Figure 16-3). For example, if the angle of the shoulder is 50 degrees, and the elbow is 90 degrees, we can use kinematics to calculate the position of the end effector using these values.

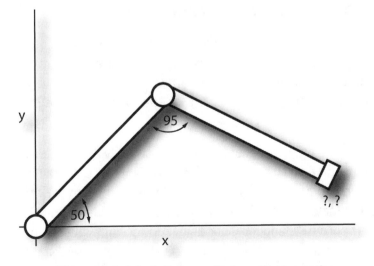

Figure 16-3: Calculating coordinates with kinematics

The other sub-branch is *inverse kinematics*, which can calculate how much to rotate the motors at each joint in order to place the hand at a specific coordinate in 2D or 3D space (see Figure 16-4). Inverse kinematics is the branch we are most interested in for this chapter and the next.

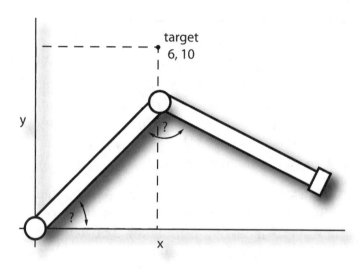

Figure 16-4: Calculating joint angles with inverse kinematics

Calculating Inverse Kinematics

The real power of a robot arm is in the code. The control software must keep track of every joint in the arm in order to calculate movements in 3D space. The angles are calculated using trigonometry, a high-school level discipline that people tend to forget with disuse.

Our robot will have three joints, but let's simplify the problem by examining an arm that has one shoulder joint and one arm joint. Our main programming goal is to command the arm to move the end effector to specific x and y coordinates—we give the program a coordinate and it responds by moving the end effector to that coordinate. The position of the end effector is determined by two things: the angle of the shoulder and the angle of the elbow.

To restate the problem, we give it two numbers (x and y co-ordinates), and it uses these numbers to calculate two new numbers (shoulder angle and elbow angle). In order to do this, we will use geometry and triangle math.

We start with x and y, which is the target coordinate, and the length of each segment of the arm. However, we don't know the angles we need to rotate the motors to in order to make the arm move to these coordinates. It helps to visualize where the robot arm should be (see Figure 16-5). In this view we can clearly see that the arm makes up a triangle consisting of three points: motor 1, motor 2, and the end effector. The obvious triangle geometry is why we use triangle math to determine the angles for the two motors.

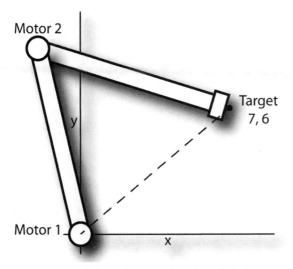

Figure 16-5: Imagining a triangle within the arm

Each side of the triangle has a length. Two lengths are known (the length of the forearm and main arm) and the final length is unknown. We will label these a, b, and c (see Figure 16-6). The angle between a and b is labeled A. The *law of cosines* has the following equation as it applies to triangles:

$$c^2 = a^2 + b^2 - 2ab\ \cos A$$

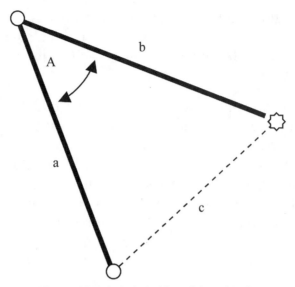

Figure 16-6: Labeled sides of the triangle

If we can determine the distance of c we would know every variable in this equation needed to calculate A. So how do we find c? Since we know the coordinate values of x and y (for example, let's use x=7, y=6 as coordinates), we can create a right-angled triangle that includes c (see Figure 16-7). Now that we have a right angled triangle, we can calculate c using:

$$c^2 = a^2 + b^2 \quad \text{Rewritten as...} \quad c = \sqrt{a^2 + b^2}$$

Now that we know a, b, and c we can revisit our first equation. We want to calculate A, so we rewrite the equation as follows:

$$A = \mathrm{acos}\left(\frac{a^2 + b^2 - c^2}{2ab}\right)$$

I won't bore you with the equations for the base angle where the main arm rotates, since it is similar to the other angle calculation. All you need to know is that the base angle is determined by adding B1 and B2 (see Figure 16-8), since the arm will start at angle 0, which points East in a Cartesian coordinate system.

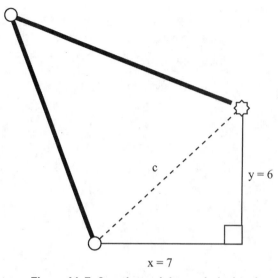

Figure 16-7: Creating a right-angled triangle

That's all there is to it! As you can see, the math to calculate inverse kinematics is not too difficult. In the next chapter we will put these equations into practice by building and programming a robot arm to navigate in three dimensional space.

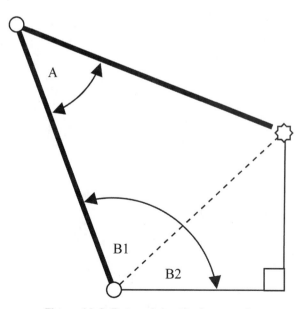

Figure 16-8: Determining the base angle

CHAPTER 17

Robotic Arm

TOPICS IN THIS CHAPTER

- Building LARA
- Auto Calibration
- Programming LARA

The arm in this chapter uses three motors to move the end effector along all three axes of x, y, and z (see Figure 16-1). It is named LARA, which stands for LEGO Automated Robotic Arm. LARA is easy to build and requires few gears, mostly because of the excellent NXT servo motors. The goals for this project are as follows.

- Fast–able to move quickly
- Powerful–able to lift light masses
- Far Reaching—long appendages
- Strong–parts won't break off or fall apart
- Reliable–cord won't tangle up

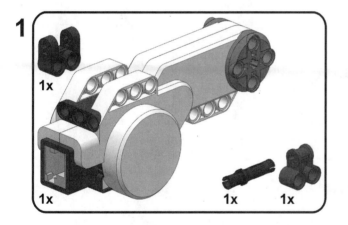

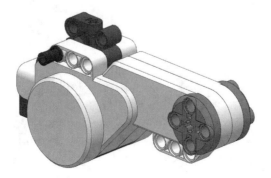

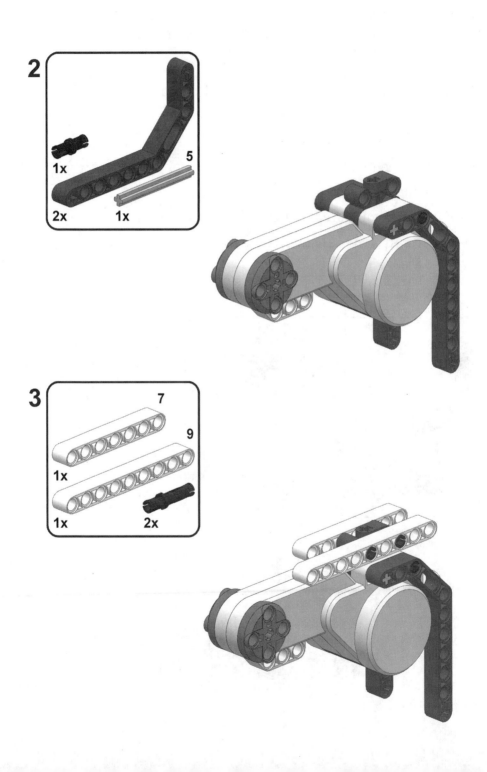

2

1x
2x   1x   5

3

7
9
1x
1x   2x

4

6x
5
2x

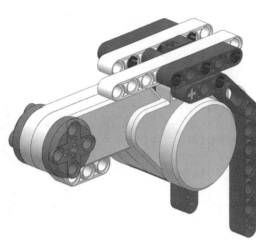

5

3
1x
1x
2x

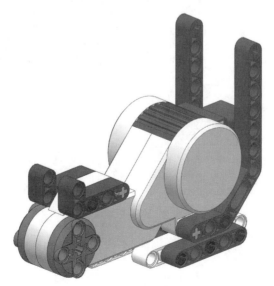

6

1x
1x
1x

7

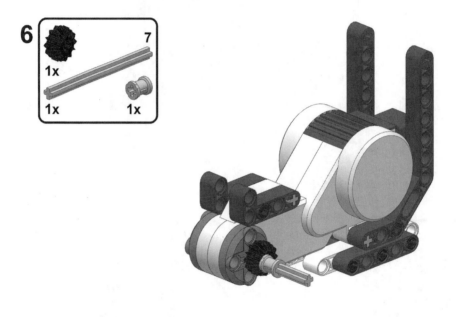

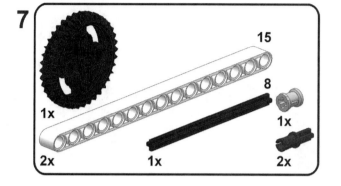

7

15

8

1x

2x   1x   1x   2x

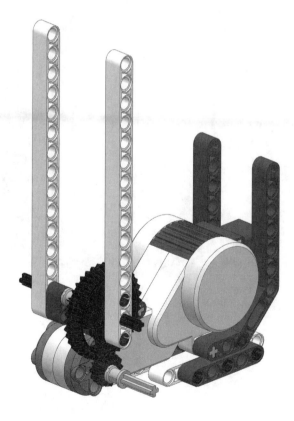

8

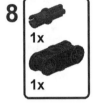

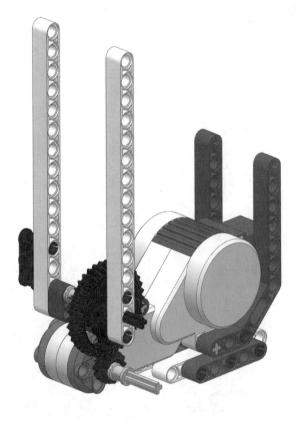

9

5

1x

1x

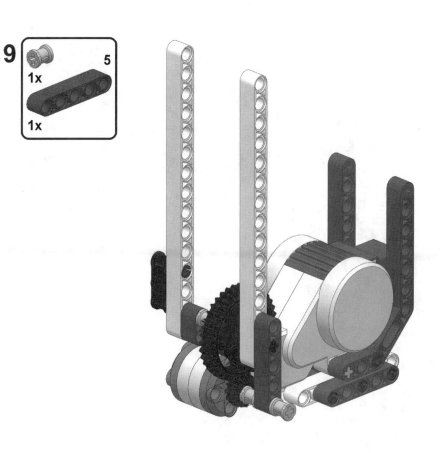

10

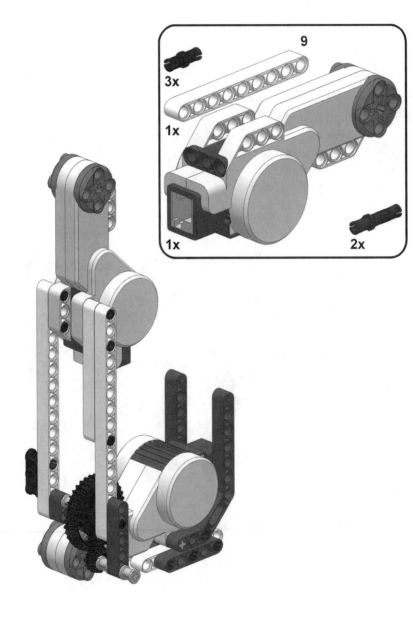

11   13

4x   8x

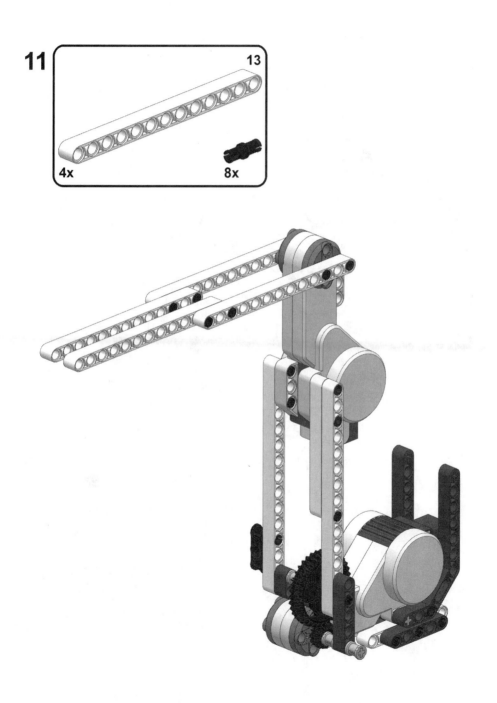

12

1x   2x

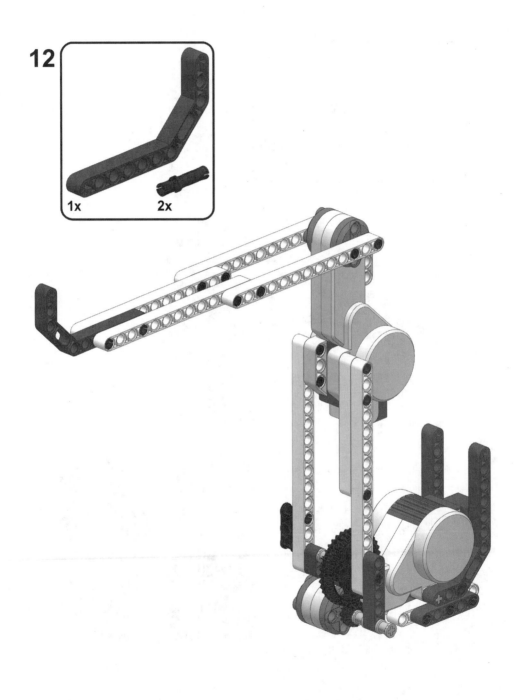

13

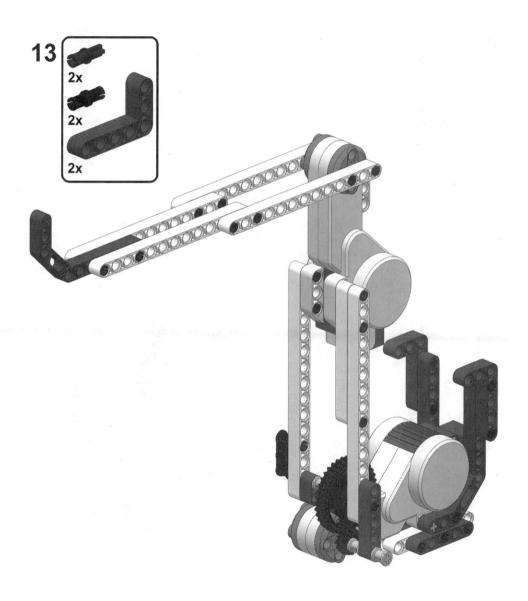

14

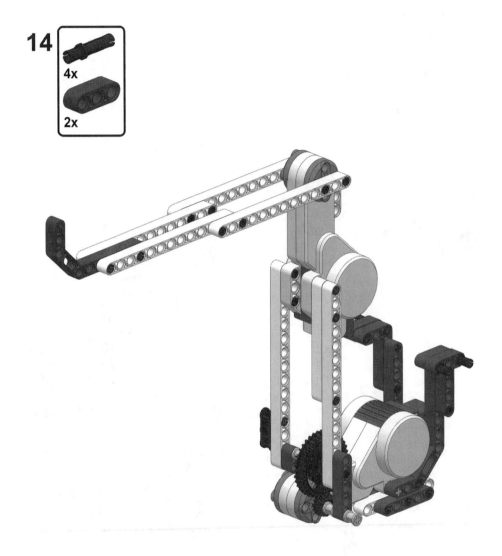

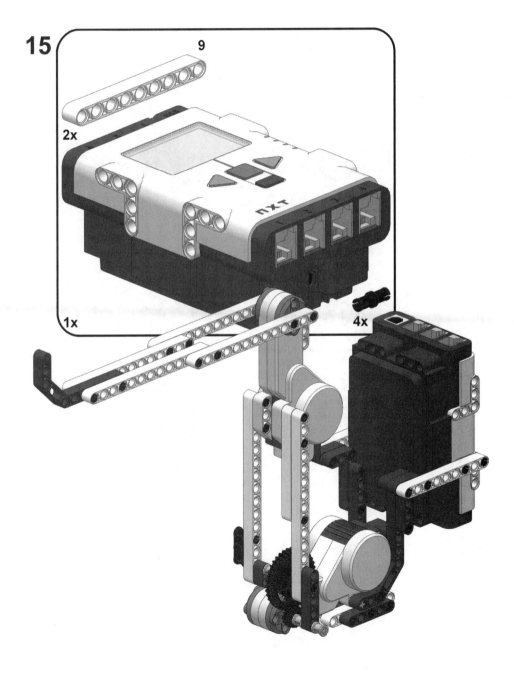

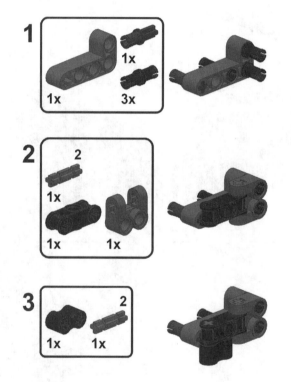

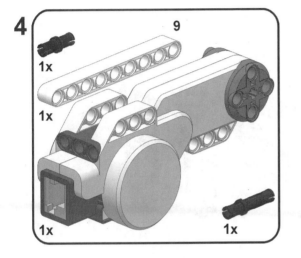

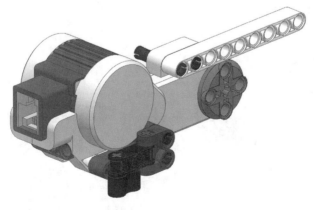

5

1x   6

1x   2x

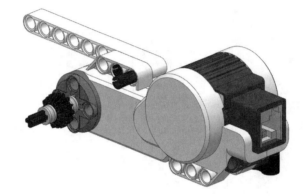

6

2x

1x

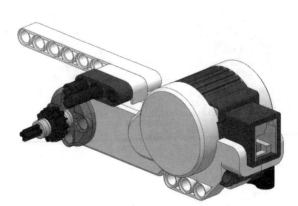

7

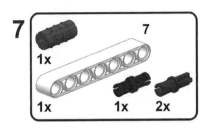

8

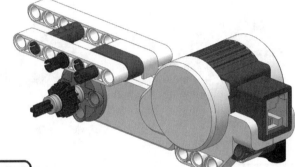

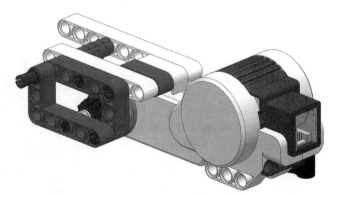

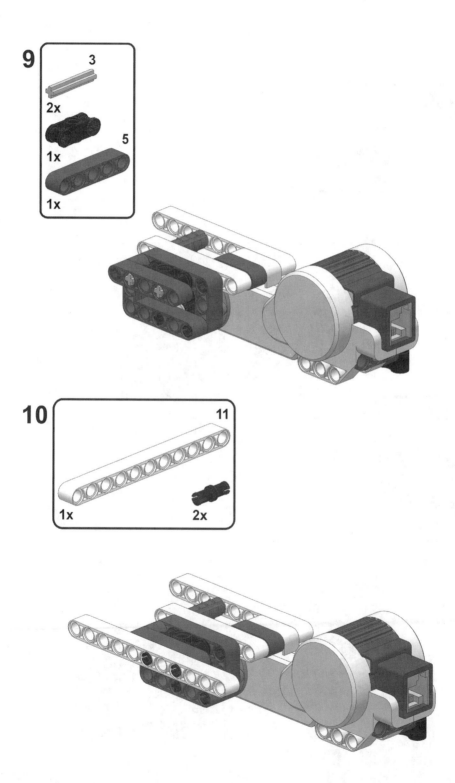

11

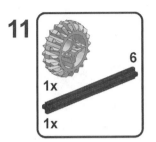

1x
6
1x

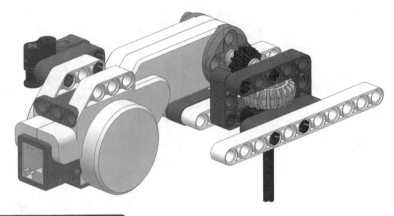

12

4x
2x   1x

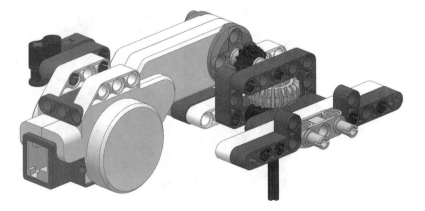

13

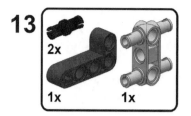

2x
1x   1x

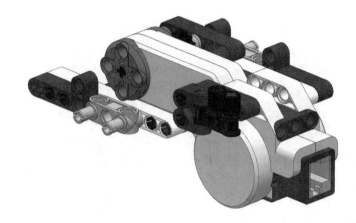

14   11

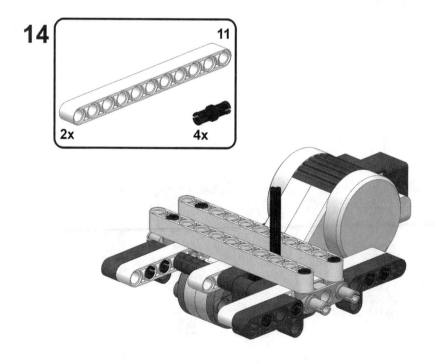

2x   4x

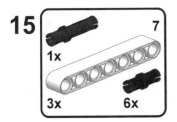

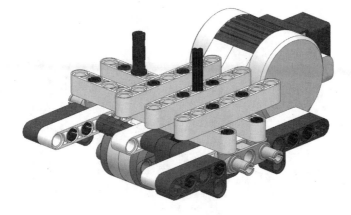

16
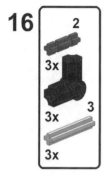

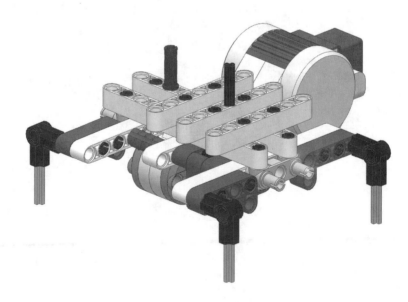

17

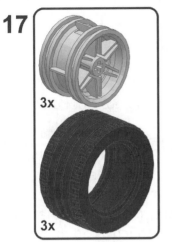

3x

3x

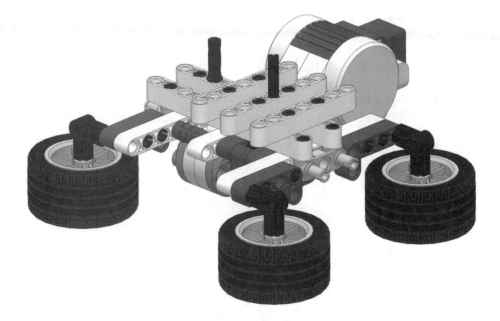

Add the arm to the base as shown in figure 16-1. Now it's time to connect the motors:

- Connect a long cable from the base motor to port A.
- Connect a short cable from the shoulder motor to port B.
- Connect a medium cable from port C to the elbow motor.

Make sure all the cables are loose and not wound around the robot.

You might notice this arm is unable to rotate a full 360 degree turn due to the cable leading to the base motor. This was a deliberate design because in order to automatically calibrate the motor tachometers (more on this later) it needs to have a stopper to limit how much it can rotate (the double-pin sticking up from the base). With this physical limiter in place, there was no reason to construct an arm able to spin all the way around.

Programming LARA

There are many similarities between the navigation classes that deal with 2D space and the task of moving an arm—navigating through 3D space. In fact, even the triangle math is similar.

The triangle math must determine three angles as opposed to the two angles described in the previous chapter. Figure 17-1 shows the top view of the arm. With this view, rotation only seems to occur around the z-axis because the elbow appears straight due to the orientation of the elbow motor. A1 is the only angle in this view.

Figure 17-2 shows the side view. From this angle, we can see two joints: A2, the shoulder joint, and A3, the elbow joint. The complete 3D coordinates are calculated using all three of these angles (A1, A2 and A3).

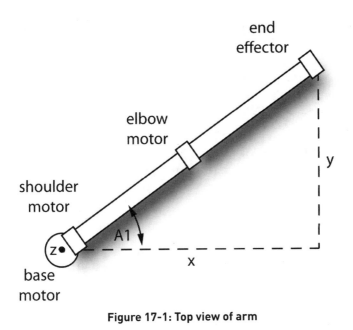

Figure 17-1: Top view of arm

Because we discussed triangle math so extensively in the earlier section we can skip this explanation and go straight to the code. There's really not a lot of code considering the complexity of the calculations involved. Much of the code has to do with calibrating the arm so you don't have to manually line up the position each time you run the program.

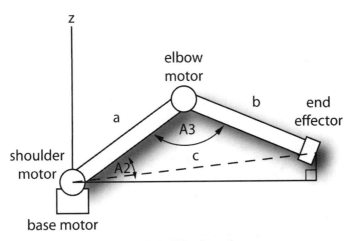

Figure 17-2: Side view of arm

```java
import lejos.nxt.*;
import lejos.robotics.RegulatedMotor;

public class LARA {
  double BASE_ARM_LENGTH = 16.3; // cm's
  double FOREARM_LENGTH = 19.3;
  double DISTANCE_FROM_CENTER = 6.5; // dist
from center axis

  int ZAXIS_START_ANGLE = -90;
  int SHOULDER_START_ANGLE = 90;
  int ELBOW_START_ANGLE = 90;

    RegulatedMotor R;
    RegulatedMotor S;
    RegulatedMotor E;

  public static void main(String [] args)
throws Exception{
    LARA a = new LARA();

    // Pick up part:
    a.gotoPoint(0, 25, -7);
    a.gotoPoint(0, 35, -5);
    a.gotoPoint(0, 50, 10);

    a.gotoPoint(-15, 10, -30); // drop part

    a.gotoPoint(-15, 10, 30); // return
    a.gotoPoint(0, 40, 15);
  }

  public LARA() {
    S = calibrate(MotorPort.B, false, -115);
// Shoulder
    E = calibrate(MotorPort.C, false, -81); //
Elbow
    R = calibrate(MotorPort.A, true, 160); //
Rotator

      R.setSpeed(400); // Rotator
    R.setAcceleration(200);
    S.setSpeed(400); //Shoulder
    S.setAcceleration(200);
    E.setSpeed(400); // Elbow
    E.setAcceleration(200);
  }

  public RegulatedMotor calibrate(MotorPort
port, boolean reverse, int target) {
    NXTMotor motor = new NXTMotor(port);
```

```
    motor.setPower(20);
    if(reverse)
      motor.backward();
    else
      motor.forward();
    int old = -999999;
      while(motor.getTachoCount() != old) {
        old = motor.getTachoCount();
        try {
          Thread.sleep(500);
        } catch (InterruptedException e) {
        }
      }
    RegulatedMotor reg_motor = new
    NXTRegulatedMotor(port);
     reg_motor.setSpeed(150);
     reg_motor.resetTachoCount();
     reg_motor.rotate(target);
     reg_motor.resetTachoCount();
      return reg_motor;
    }

    public void gotoPoint(double x, double y,
double z) {
      // 1. Calculate Z-Axis angle on shoulder
      double zaxisangle = Math.atan2(-y,x); //
-y because motor is reverse?

      // 2. Calculate shoulder angle:
      double Z = Math.sqrt(y * y + x * x);
      Z = Z - DISTANCE_FROM_CENTER; //
Corrected due to arm construction
      double c = Math.sqrt(Z * Z + z * z);
      double angle1 = Math.asin(z/c);
      double angle2 = Math.acos((BASE_ARM_
LENGTH * BASE_ARM_LENGTH + c * c - FOREARM_
LENGTH * FOREARM_LENGTH)/(2*BASE_ARM_
LENGTH*c));
      double shoulderangle = angle1 + angle2;

      // 3. Calculate elbow angle:
      double elbowangle = Math.acos((Math.
pow(BASE_ARM_LENGTH, 2) + Math.pow(FOREARM_
LENGTH, 2) - Math.pow(c, 2))/(2*BASE_ARM_
LENGTH*FOREARM_LENGTH));

      rotateZAxisTo(Math.
toDegrees(zaxisangle));
      rotateShoulderTo(Math.
toDegrees(shoulderangle));
```

```
      rotateElbowTo(Math.
toDegrees(elbowangle));

      while(R.isMoving() || S.isMoving() ||
E.isMoving()) {
        Thread.yield();
      }

      Sound.beep();
      try {
        Thread.sleep(1000);
      } catch (InterruptedException e) {}
  }

  /**
    * Rotate elbow up or down.
    * @param angle +ve value is up.
    */

  public void rotateElbowTo(double angle) {
    angle = angle - ELBOW_START_ANGLE;
    E.rotateTo((int)-angle, true);
  }

  /**
    * Rotate shoulder up or down.
    * @param angle +ve value is up.
    */

  public void rotateShoulderTo(double angle)
{
      double ratio = 36F/12; // Gear ratio
36:12
      angle = angle - SHOULDER_START_ANGLE;
      S.rotateTo((int)(angle * ratio), true);
  }

  /**
    * Rotate base of arm.
    * @param angle +ve value is
counterclockwise
    */
  public void rotateZAxisTo(double angle) {
      double ratio = 20F/12; // Gear ratio
20:12
      angle = angle - ZAXIS_START_ANGLE;
      R.rotateTo((int)(angle * ratio), true);
  }
}
```

Auto Calibration

It is important to know the position of each segment of the arm when the program starts, otherwise it will be confused about where it is moving the end effector. The calibration method works by setting the motor power very low, then rotating each part of the arm until the motor seizes. At that point it knows the arm has hit itself and it records the position of the arm. It then moves back to the starting position and resets the motor tachometers to zero.

Using the Robotic Arm

The program above takes the robot arm through a designated path. It will try to hook a plastic LEGO ring and dump it over the edge of your table. To use this program, assemble the plastic ring as shown in Figure 17-3. Then place the arm on a table and line it up along the edge as shown in Figure 17-4.

Figure 17-3: Assembling a simple LEGO ring

When the program starts, LARA will automatically go through the calibration routine. After this is done, it will move the arm into the neutral position. This looks something like a cobra ready to strike.

It will then pick up the ring and dip the arm below the table in the negative z-axis quadrant in order to let the ring slip off the hook.

The motor control algorithms help the arm maintain a controlled speed. Hopefully this project reminds you of the way industrial robots move (though probably a little less precise).

The origin point (0, 0, 0) is located above where the axle meets the base. It is at the same height as the large black gear at the base of the arm, just above LEGO insignia on the shoulder motor (see Figure 17-5). Place the ring about 25 centimeters in front of the origin point..

Just for fun!

LARA would be ideal for remote control from your PC using either a USB cable or Bluetooth communications. The next few chapters discuss PC to NXT communications. This will give you the knowledge necessary to adapt the LARA code so that you can input coordinates through the PC keyboard. It would be much faster to set up rings and command the arm to pick them up and drop them off.

This was just a demonstration, but now it is up to you to change the code in the main() method to make the robot perform other tasks. Try building some more rings and placing them at precise positions around the robot arm. See if your code can guide the hook to pick up multiple rings and dump them over the edge.

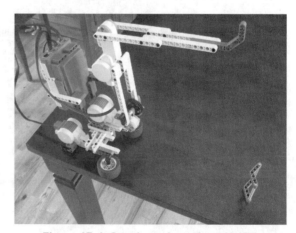

Figure 17-4: Starting orientation of LARA

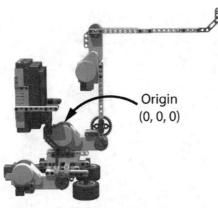

Origin
(0, 0, 0)

Figure 17-5: Locating the origin point

Illegal Moves

One topic not addressed by the code is the possibility of an illegal move. Since we are navigating through 3D space, much like a mobile robot, there are obstacles that the arm cannot pass through, such as the table or space occupied by the robot arm itself.

The arm has a limited length. If you ask it to move the end effector to a point beyond the reach of the arm, it should recognize this situation and elegantly back out of such a move. Furthermore, the arm cannot reach back to its own origin point (0, 0, 0) because that is where the motor and NXT brick occupy space. However, in-depth heuristics are beyond the scope of this chapter and would make the code even longer than it already is. Feel free to take this programming challenge yourself and improve the code.

Just for fun!

Try adding the ultrasonic sensor to your robot arm to see if it can locate rings and attempt to hook them. Or, if you have an NXT cam (see chapter 24), try adding vision to LARA. The NXT cam is capable of identifying objects by color and locating x and y coordinates within the field of view.

CHAPTER 18

Remote Communications

TOPICS IN THIS CHAPTER

- Bluetooth Introduction
- PC Comms
- Starting an Eclipse Project
- PCComms API

Wireless communication is invaluable to robotics be-cause of the enormous resources and that can be harnessed. Using Bluetooth, you can use the superior pro-cessor and memory of your PC to control robots. Bluetooth has even been used to control prosthetic legs that allow double-amputees to walk again. The possibilities are end-less. In this chapter we'll find out how to easily control a robot from your PC.

Bluetooth Communications

Bluetooth allows you to make a wireless connection be-tween two devices, almost like a USB port without cables. It is ideal for computer peripherals like keyboards and mice, but it also has many other uses.

Bluetooth (and other wireless protocols) create a *Wireless Personal Area Network* (WPAN), which is almost like a por-table wireless network that follows you around wherever you carry your Bluetooth enabled device. The Bluetooth adapter is actually a wireless hub.

To use Bluetooth, you need a Bluetooth adapter (see Figure 18-1). The NXT kit does not include a adapter since Blue-tooth is not required to upload programs to the NXT. Do you really need one? The short answer is yes, go right now and purchase a Bluetooth adapter. You can find Bluetooth adapters in electronics stores, from LEGO, and on eBay.

> **WARNING:** Make sure to buy a Bluetooth 2.0 adapter as the NXT does not work with Bluetooth 1.2 and older versions. Versions higher than 2.0 will also work, because Bluetooth is backward compatible.

Figure 18-1: Bluetooth Adapter

Not all Bluetooth adapters are compatible with the NXT. First, it must be Bluetooth 2.0 or higher. Second, under Windows it must support the Widcomm® Bluetooth stack. Some Bluetooth solutions, such as those built into Dell notebook computers, are not compatible. If you are not sure and want to avoid researching this, you can order a Bluetooth adapter from LEGO. As long as Linux users can use their adapter with the Bluecove Bluetooth stack, there should be no compatibility problems.

If you want to confirm that your adapter is using the Widcomm Bluetooth stack, do the following:

1. Select Start > Control Panel > System.

2. Select the Hardware tab, then click the Device Manager button.

3. You will see a list of devices, including Bluetooth. Expand the Bluetooth selection and highlight the Bluetooth wireless hub (see Figure 18-2).

4. Click the Properties icon and you will see some general information. Click the driver tab to view the device driver (see Figure 18-3).

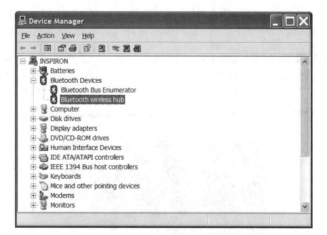

Figure 18-2: Bluetooth devices in Windows

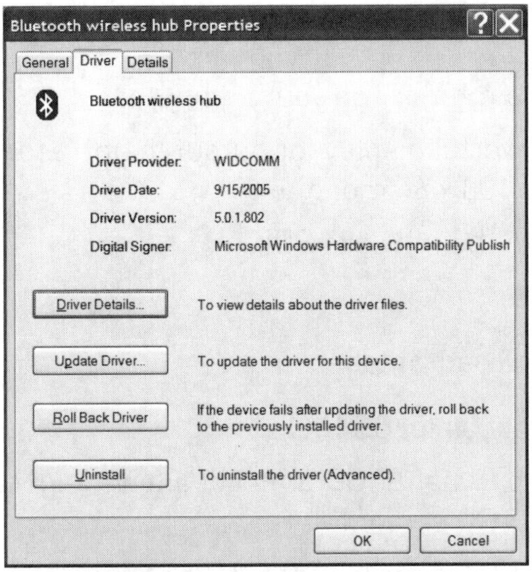

Figure 18-3: Device driver information

As long as you see Widcomm and any version higher than 1.4.2.10 you should have no compatibility problems.

PC Comms

Throughout the first part of this book we ran code on the NXT brick. But there is another mirror-image of the NXJ API that runs on the PC called PC Comms. This package contains almost every class in NXJ and a few more classes not available for the NXT. The package seeks to mirror leJOS NXJ as closely as possible, which means the classes and methods tend to look similar.

The major difference is that leJOS NXJ code runs on the NXT brick while PC Comms code runs on your computer. It controls the NXT brick by sending individual commands wirelessly. Whether you choose to use leJOS NXJ or PC Comms depends on the requirements of your robot.

PC Comms Advantages:

• Access to all devices on your computer (see Figure 18-4)

- Access to memory on your computer.
- More powerful display for output (high-resolution full color vs. 100 x 64 mono)
- Access to the full Java SDK

PC Comms Disadvantages:

- Not as fast as leJOS NXJ
- Limited by what LEGO protocol allows (e.g. cannot draw on LCD screen)
- Not as portable if you want to take your robot somewhere.

Speed is often a concern for robot programmers. Because all commands are sent through Bluetooth, commands take a little longer to reach the NXT brick. However, most robots are slow and do not require fast processors. It takes a few seconds to move a foot, so a hundred milliseconds here and there isn't a big deal for most robot applications.

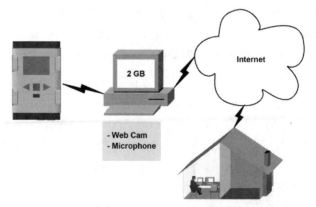

Figure 18-4: Exploiting your computer's resources

Since PC Comms is running on your computer and uses the standard Java API, you have access to an incredible number of resources, such as the Internet and other hardware devices on your computer. Your imagination can go wild thinking of interesting new creations.

Memory is limited on the NXT brick. Most projects in this book can reside wholly in NXT brick memory. However, if you want to create a super project that incorporates many projects in this book into one robot (for example, a mobile robot that maps the environment automatically and also has a robotic arm to transport objects) then you would have a problem fitting everything on the NXT. Using PC Comms you can amalgamate many of these complex programs without taxing the gigabytes worth of memory on your computer.

Pairing the NXT with a PC

Before you can use Bluetooth you must pair the NXT brick with your PC. Pairing occurs when one device tells another device that they can be friends. The reason for pairing is to prevent strangers from accessing your Bluetooth devices without permission. By pairing, you give the devices permission to interact (like plugging a USB cable into a computer).

Pairing is done by locating the Bluetooth device, then entering a four digit code on both devices. I recommend using the LEGO software to pair the devices. If there are any problems, the software will attempt to help you.

If you don't want to use the LEGO software, you can pair the devices using your operating system.

1. Turn on your NXT brick. From the main menu, select Bluetooth (see Figure 18-5).

Figure 18-5: Selecting the Bluetooth menu

2. Hit the orange button again to activate Bluetooth if it is not already on (see Figure 18-6).

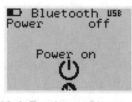

Figure 18-6: Turning on Bluetooth power

3. When power is on you will see a Bluetooth icon in the upper right corner and several menu options (see Figure 18-7). If visibility is off, select the visibility icon (an eye) and press the orange button.

Figure 18-7: Turning on visibility

4. Now that the NXT brick is visible, you can begin the pairing process. In Windows, select Start > Control Panel > Hardware and Sound. Before going any further, make sure your NXT brick is turned on. Below Bluetooth Devices is the option to Add a Wireless Device. Click this.

5. You should see a list of Bluetooth devices discovered by your PC. Likely you will just see one in the list (see Figure 18-8). Select this item and click next.

6. After a while your computer will prompt you to enter a code (see Figure 18-9).

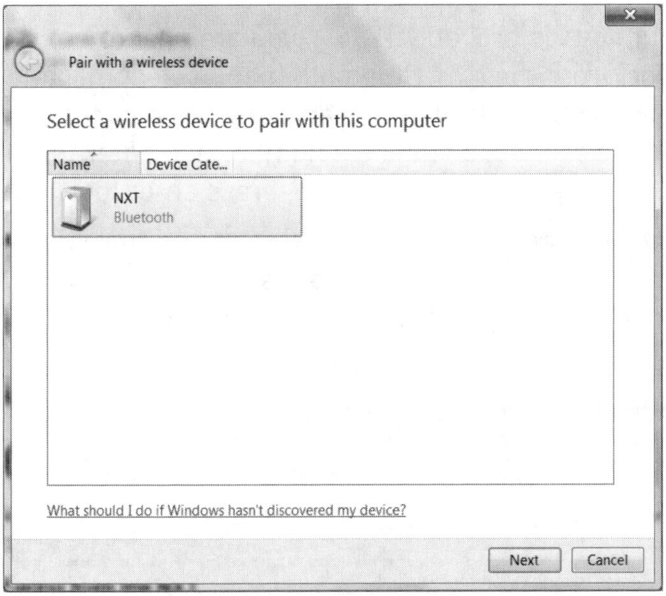

Figure 18-8: Discovering Bluetooth devices

Figure 18-9: Entering the PIN code

7. Your code is probably 1234. If you are unsure of the code, from the NXT Bluetooth menu select PIN and it will display the code. You can change the code by pressing the arrows to change a number and enter to select it. Or to keep the default code, press enter four times.

8. Once you enter the correct PIN code on the PC it will complete the pairing process (see Figure 18-10). Click close and you are ready to communicate.

Figure 18-10: Successfully pairing the NXT brick

9. Now that your NXT is paired with your PC, you can upload code to the NXT brick via Bluetooth! In Eclipse, select Help > Preferences > leJOS NXJ. Click the radio button next to Bluetooth (see Figure 18-11). Optionally you can also enter the name of your brick if you want to make sure other bricks are not used. The name is located at the top of the NXJ main menu. Click OK.

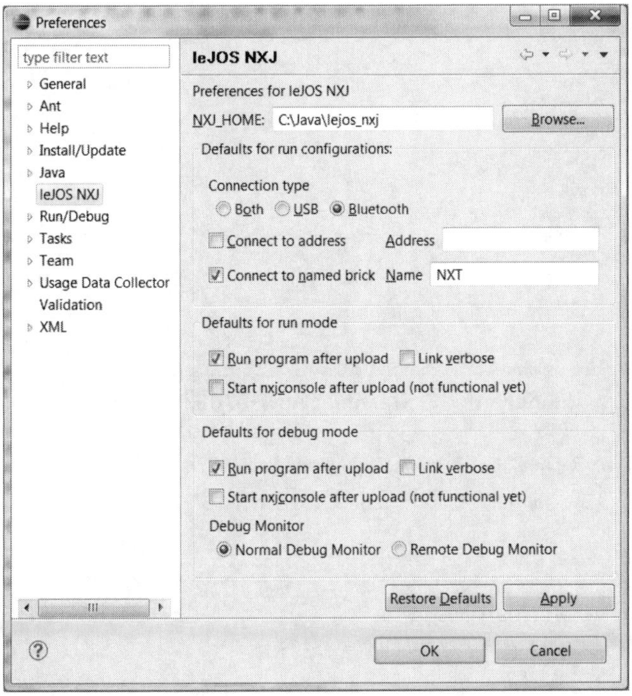

Figure 18-11: Selecting Bluetooth in Eclipse

Now when you click the green run button, Eclipse will attempt to upload your code via Bluetooth.

Starting an Eclipse PC Project

The easiest way to program PC Comms code is through the Eclipse leJOS Plugin. Assuming you have already installed Eclipse (see Chapter 2), all you need to do is start a new leJOS PC Project.

1. In Eclipse, select File > New > Project...

2. Under Lejos, select Lejos PC Project and click next (see Figure 18-12).

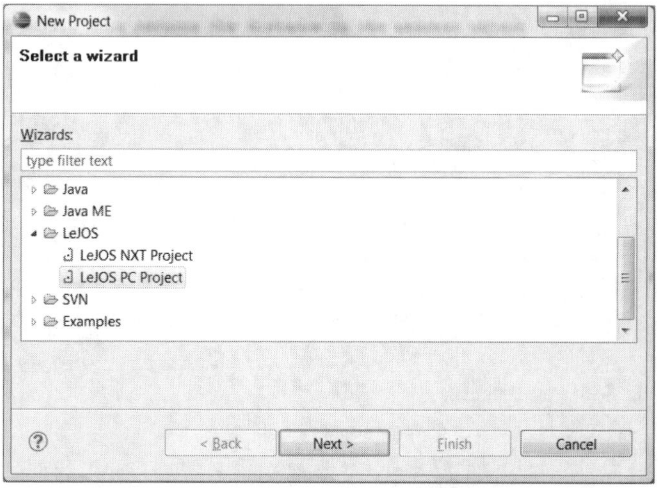

Figure 18-12: Starting a new leJOS PC Project

3. Enter a name for your project, such as PC leJOS Projects (see Figure 18-13). Click Finish.

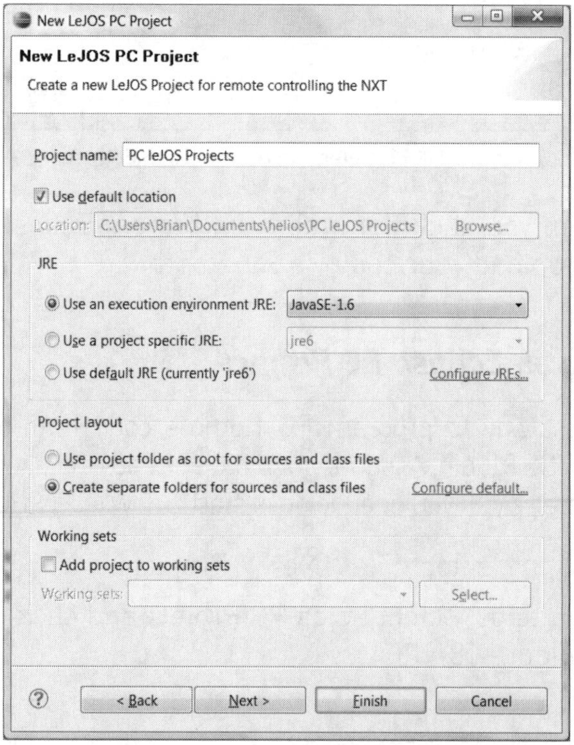

Figure 18-13: Naming the PC project

4. Now that the project is set up, you can start a new class as usual by selecting File > New > Class.

5. You can now write a program for your NXT as you normally would. For example, try a one-line program to rotate a motor 360 degrees.

6. When your code is done, turn on the NXT and hit the green run button. In a few seconds the NXT will connect and the motor will rotate.

That's all for setup. There are a few more things to learn about using PC Comms properly, which will be covered later in this chapter.

Command Line Setup

The steps below describe how to set up PC Comms if you are programming from a command line. This setup assumes you have already installed leJOS for the command line from Chapter 1.

1. Make sure you have the latest version of the Java Development Kit installed. If you installed leJOS, you already have this installed.

2. Add the PC Comms library to your classpath.

   `%NXJ_HOME%/lib/pc/pccomm.jar`

3. Add the Bluetooth library to your classpath.

   `%NXJ_HOME%/lib/pc/bluecove.jar.`

4. Remove classes.jar from the classpath, otherwise this will conflict with pccomm.jar (they contain the same package names).

Step four is a good reason to start using an IDE which accepts multiple environment settings, such as Eclipse.

Testing PC Comms

That's it. You are now ready to use PC Comms. You can use either the standard LEGO firmware or the leJOS replacement firmware. PC Comms works with both because it uses the LEGO Communications Protocol (LCP).

There are several sample programs included with PC Comms. To test these, compile and run them as normal Java programs. They will compile and execute within Eclipse without any special development tools.

If you are using the command line to compile and run Java code, use the regular JAVAC and JAVA commands. For example, in the PCSamples directory is a program called SensorTest.

```
javac SensorTest.java
```

Now run the program:

```
java SensorTest
```

With SensorTest, hook up some sensors to the NXT, and when you run the program you should see the values output to the console. This means everything is installed and working properly.

PC Comms API

As previously mentioned, the PC Comms API is similar to leJOS NXJ. In fact, it is almost identical. In some places such as the navigator, behavior and sensor classes, the source code is the same. It would be redundant to list the entire PC Comms API, but you can browse the classes from the leJOS website at www.lejos.org (see Figure 18-14).

The most important difference between PC Comms and leJOS NXJ is that PC Comms takes a few seconds to establish a connection with the PC. This happens automatically the first time you call a command, such as a motor command. It also automatically closes the connection when the program ends.

Note: Motor rotations are more accurate when you use the leJOS firmware as opposed to the standard LEGO firmware. If you decide to use PC Comms with the LEGO firmware, make sure to upgrade to the latest version of the firmware.

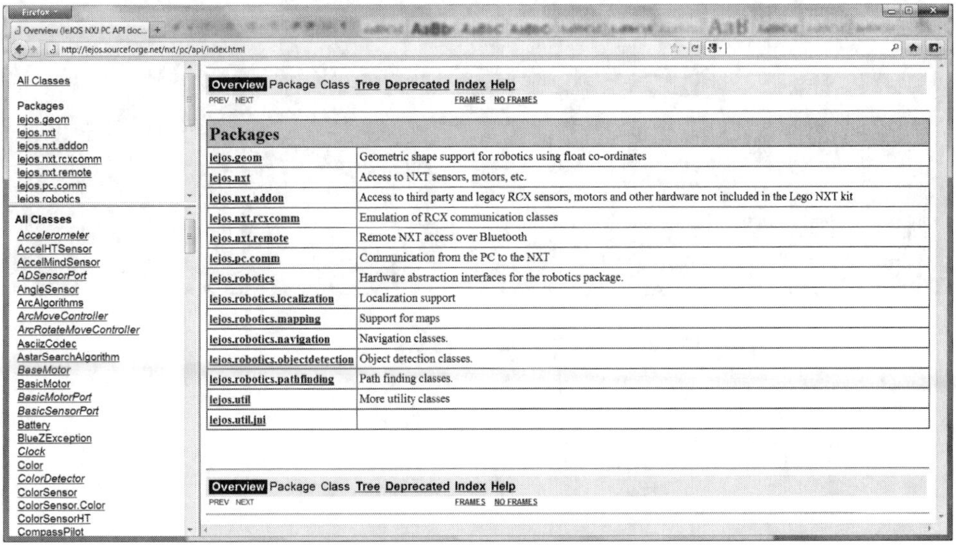

Figure 18-14: The PC Comms Javadocs

CHAPTER 19

Remote Control Vehicle

TOPICS IN THIS CHAPTER

- Building Speed Buggy
- Programming Remote Control

One of the classic uses of radio communications is the radio controlled (RC) car. In this chapter we will build and program an RC car, or in this case, an RC dune buggy (see Figure 19-1). This little turbo charged beauty uses analog steering. I call it turbo charged because the gears speed up motor rotation 1.67 times.

Figure 19-1: The LEGO Speed Buggy

Building Speed Buggy

Speed Buggy was an early 1970s cartoon featuring a talking dune buggy with a personality. His inventor even controlled the vehicle with a remote device from time to time. With this in mind, let's build a LEGO version of Speed Buggy.

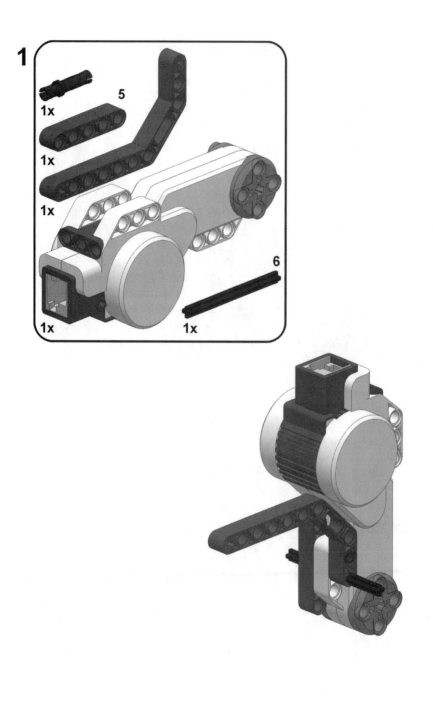

2

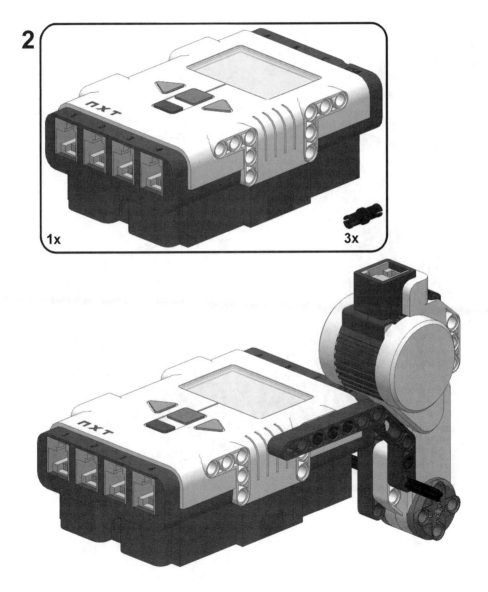

1x   3x

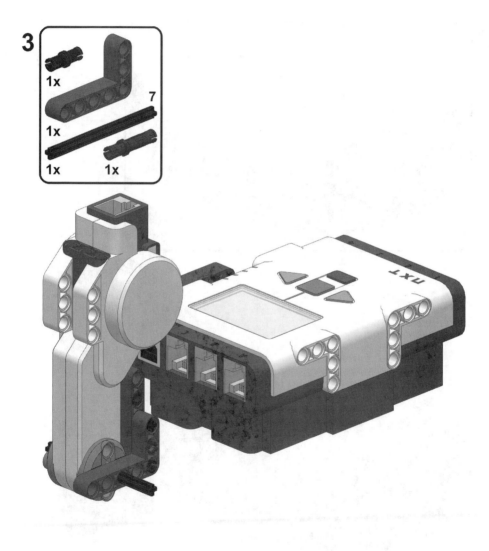

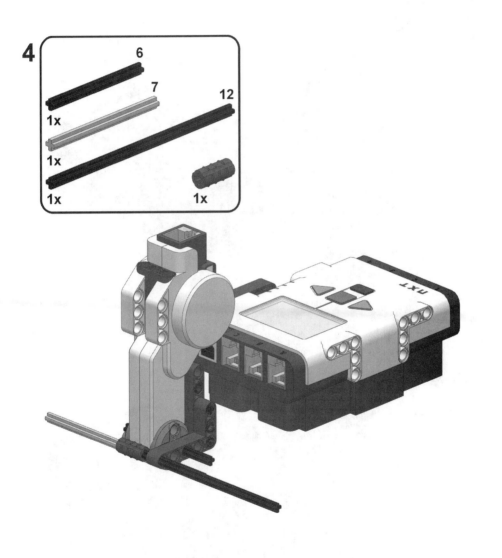

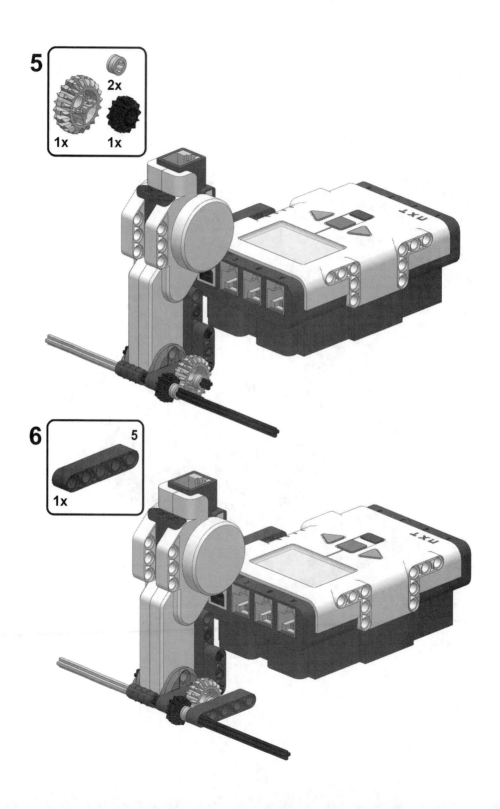

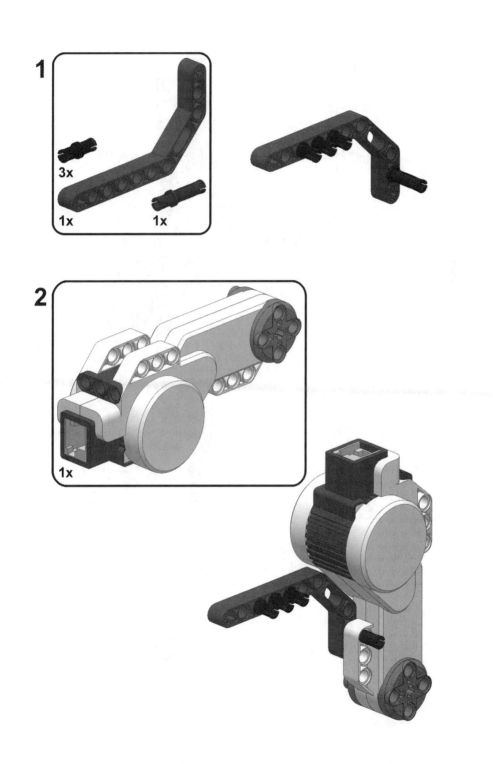

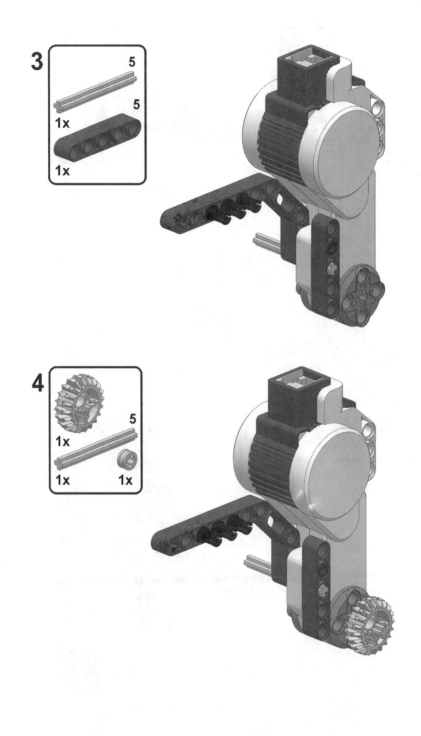

7

8

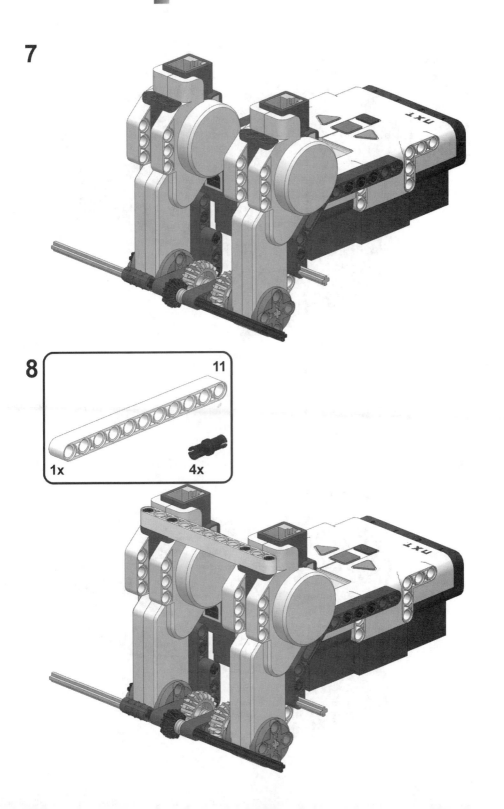

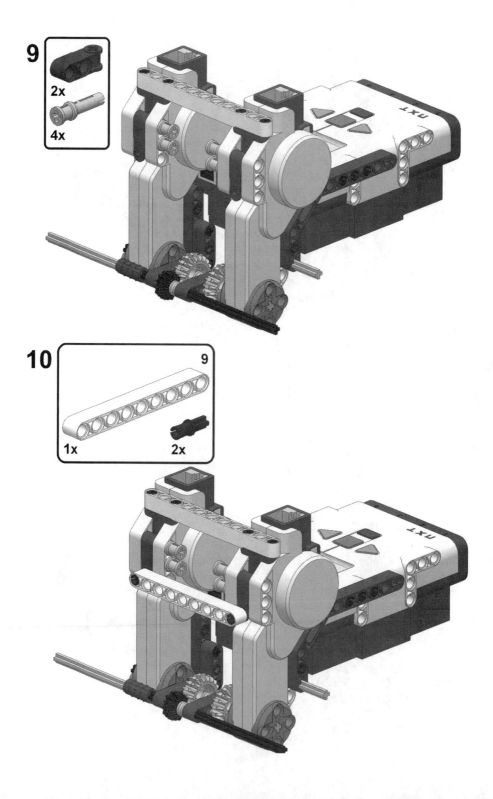

1

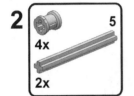

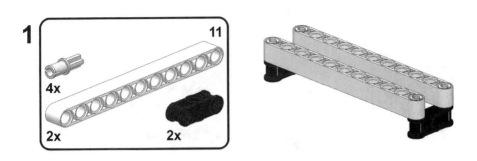

2

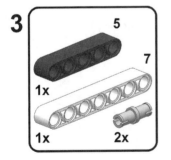

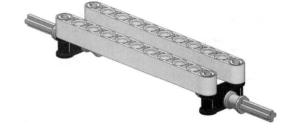

3

4

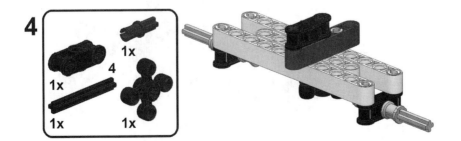

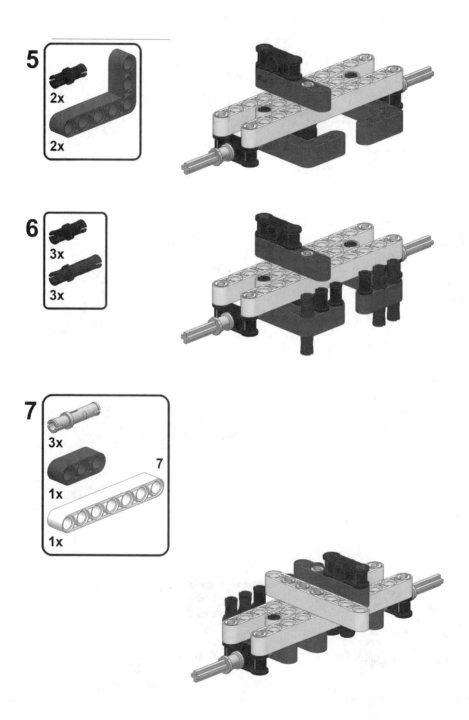

5
2x
2x

6
3x
3x

7
3x
1x
7
1x

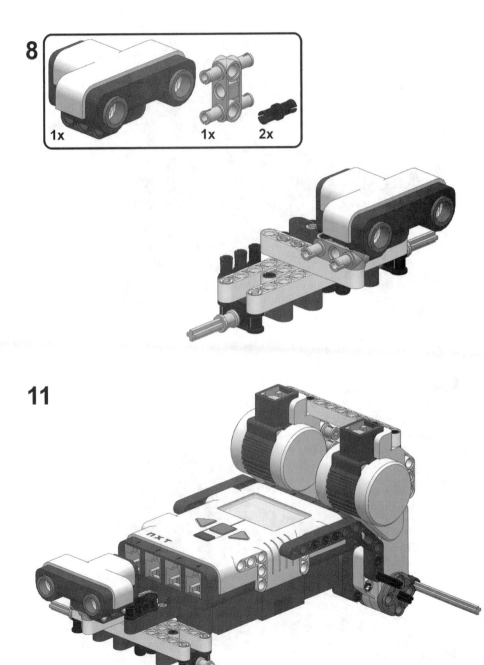

12

1x

2x

1x

13

1x

5

8

1x

1x

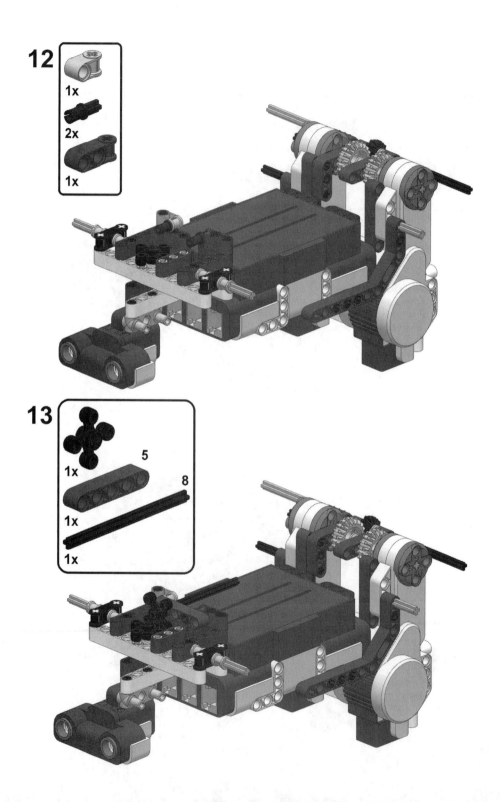

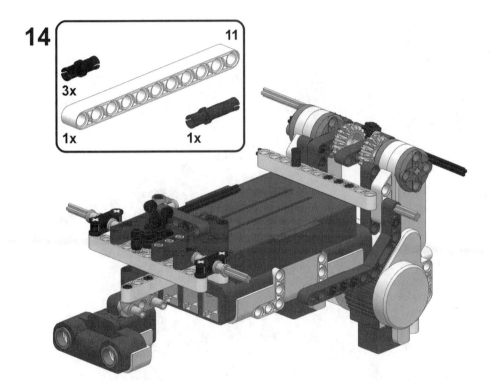

15

1x

7

1x

1x

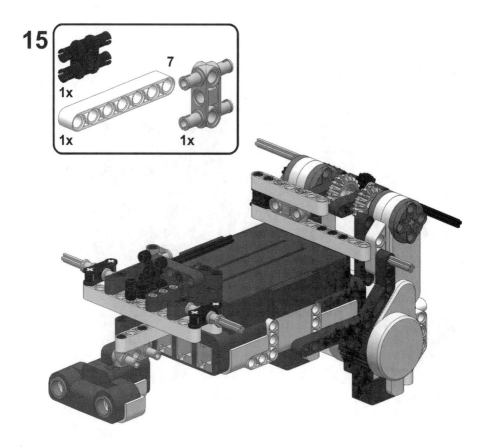

16

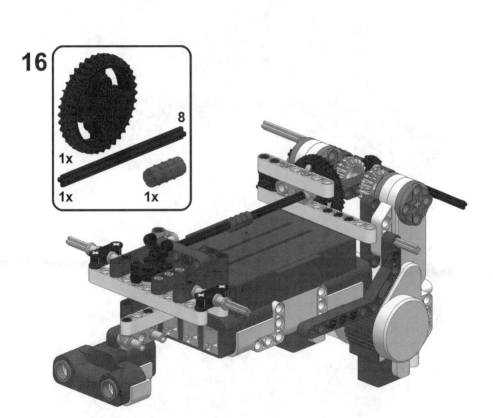

8

1x

1x   1x

17

1x

1x

18

7

1x

1x

1x

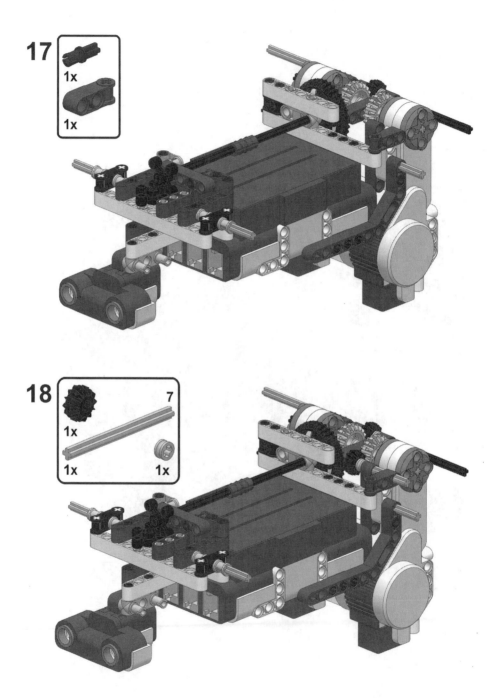

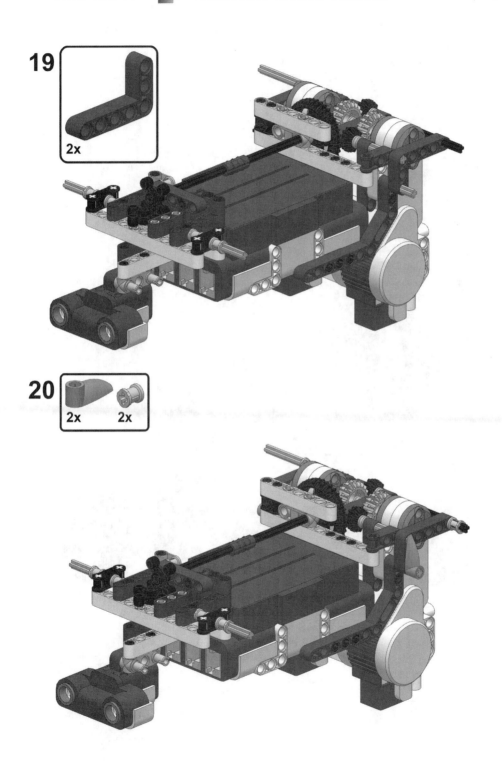

21

4x

4x

Now it's time to plug in the cables:

- Connect a short cable from the ultrasonic sensor to port 1
- Connect a medium cable from the right motor (the steering motor) to port B
- Connect a medium cable from the left motor (the drive shaft motor) to port C

Ideally, this vehicle would use a differential for the rear axle. Without the differential, both rear tires rotate at the same rate when the vehicle is turning. Something has to give, so one of the tires will skid. This also results in a sub-optimal turning radius. With a differential along the rear axle, the turning radius would decrease dramatically. The NXT kit does not contain one, so we will do without.

Programming the Radio Controlled Car

The code below allows you to control the Speed Buggy remotely using a keyboard. Within Eclipse you will need to enter this code into a leJOS PC Project (not a leJOS NXT project). Review Chapter 18 for a quick refresher on how to create, edit and run code within a leJOS PC Project.

```
import lejos.nxt.*;
import java.awt.event.*;
import java.awt.*;

public class SpeedBuggy extends Frame
implements KeyListener{

  final int FORWARD = 87, // W = forward
  BACKWARD = 83, // S = backward
  LEFT = 65, // A = left turn
  RIGHT = 68, // D = right
  QUIT = 81; // Q = quit

  static final int LEFT_TACHO = -80,
  RIGHT_TACHO = 80,
  CENTER_TACHO = 0;
```

```java
boolean steerreleased = true;
boolean drivereleased = true;

UltrasonicSensor us;

public SpeedBuggy(String title) {
  super(title);

  Motor.B.setSpeed(400); // Steering
  Motor.C.setSpeed(900); // Drive motor
  us = new UltrasonicSensor(SensorPort.S1);

  this.setBounds(0, 0, 300, 50);
    this.addKeyListener(this);
    this.setVisible(true);
}

public void keyPressed(KeyEvent e) {
  switch(e.getKeyCode()) {
  case FORWARD:
    if(drivereleased==true)
Motor.C.backward();
    drivereleased = false;
      break;
  case BACKWARD:
    if(drivereleased==true)
Motor.C.forward();
    drivereleased = false;
      break;
  case LEFT:
    if(steerreleased==true)
Motor.B.rotateTo(LEFT_TACHO);
    steerreleased = false;
      break;
  case RIGHT:
    if(steerreleased==true)
Motor.B.rotateTo(RIGHT_TACHO);
    steerreleased = false;
      break;
  }
}

public void keyReleased(KeyEvent e) {
  switch(e.getKeyCode()) {
    case FORWARD:
    case BACKWARD:
      Motor.C.flt();
      drivereleased = true;
      break;
    case LEFT:
```

```
    case RIGHT:
      Motor.B.rotateTo(CENTER_TACHO);
      steerreleased = true;
      break;
    case QUIT:
      Sound.beepSequenceUp();
      Sound.pause(1000);
      System.exit(0);
  }
    System.out.println("DIST: " +
us.getDistance());
  }

  public void keyTyped(KeyEvent
e) {}

  public static void
main(String[] args) {
    new SpeedBuggy("Enter
commands");
  }
}
```

Results

Turn on the NXT brick and then run the code. In Eclipse, simply press the green run button. The code for Speed Buggy is quite simple. It uses the java.awt.event package to listen for key presses from the keyboard. A small Java frame is generated so the Java code has somewhere to receive key events (see Figure 19-2). This window is small and lacking GUI components, so look carefully for it in the upper left corner of your monitor. It takes a few seconds to appear because Bluetooth must connect first.

Just for fun!

After you run this program, try adding code to make Speed Buggy talk by playing a sound file, just like his cartoon counterpart. The Sound. playSample() method will come in handy. You will need to upload a sound file to the NXT brick using NXJControl located in the bin directory of the leJOS install (more on this utility in chapter 12). If you are ambitious, try attaching several phrases to different keys on the keyboard.

Figure 19-2: Controlling the RC car from a Java window

Use the W and S keys to drive forward and backward. Use A and D to turn left and right. When you are done, press Q to quit. As previously mentioned, the steering radius is not sharp.

> **NOTE:** When the NXT battery charge is low, Bluetooth communication becomes unreliable. Make sure the batteries are charged.

LEGO Communication Protocol

The code for PC Comms sends commands to a server running on the NXT brick. This server runs when the menu system is running. All of this is invisible to you, but it is happening. The PC sends numbers to the NXT brick, and it interprets these numbers and performs a command depending on which numbers it receives. These commands form a protocol.

We use the *LEGO Communication Protocol*, known as LCP. This works with either the standard LEGO firmware that comes on your NXT brick by default, or it works with the leJOS NXJ firmware. The SpeedBuggy code used LCP behind the scenes, but you can also use LCP directly. These classes are in the package lejos.nxt.remote.

lejos.nxt.remote.NXTCommand

Using the NXTCommand class, you can directly send LCP commands. This class is the leJOS implementation of the LEGO Communications Protocol, which is used for communications between the PC and LEGO firmware.

Methods in this class allow you to control motors, read sensors, play sounds, access files, and more. If you want to take direct control of the robot using LCP, please consult the PC API Javadocs located on the leJOS home page. All of the public methods are available from the NXTCommand class in the lejos.nxt.remote package.

When browsing the Javadocs, you will notice there are fewer classes in the PC Comms package. This is because many of the classes in the leJOS NXJ API are recreations of standard Java classes, such as those found in the java.lang package. You still have access to these classes when using PC Comms, but the Javadocs for them are found elsewhere in the standard Javadocs.

Just for fun!

Want a challenge? Try blindly piloting Speed Buggy using limited data. First, draw a rough map of the first floor of your home, and place measurements in centimeters (see Figure 19-3). Now try driving around your home using only feedback provided by a PoseProvider (see chapter 11) and ultrasonic sensor. See if you can pilot the car to the far end of the home and back without looking at the vehicle. You can sweep the ultrasonic sensor back and forth to determine distances to objects.

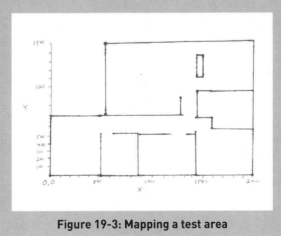

Figure 19-3: Mapping a test area

Dynamic Stability

TOPICS IN THIS CHAPTER

Gyroscopes are used to measure orientation. For example, if you rotate an object about an axis, a gyroscope mounted to the object can sense that the object has rotated. There are many uses for gyroscopes in robotics, such as sensing when a robot has fallen over. The inverted pendulum problem is one of the most exciting uses of a gyroscope, which is solved by the Segway (see Figure 20-1). In this chapter, we will build and program a mobile robot that solves this problem too.

Dynamic Stability Theory

Anything that falls over when left unsupported is inherently unstable. This includes objects like a broom sitting on your finger, a two-wheeled cart, a person doing a handstand, or a motorcycle. However, we have all seen these objects balanced. How do these objects maintain dynamic stability without falling over?

Figure 20-1: The Segway™ is an inverted pendulum

The theory behind *dynamic stability* is simple: hold a position until the object starts to fall, then shift the center of gravity to counter the direction it is tilting. Pole balancing demonstrates this concept clearly. When the pole begins to lean to the left (see Figure 20-2a), the balancer quickly moves the base of the pole to the left, which reestablishes the center of gravity (Figure 20-2b). When it starts to lean in another direction, the balancer shifts the center of gravity again.

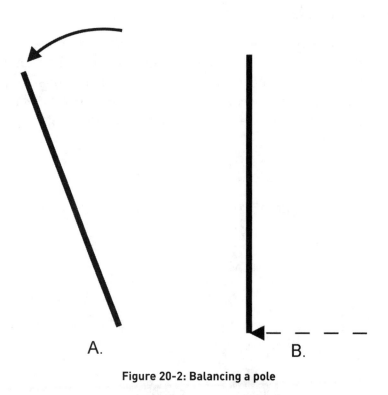

A.   B.

Figure 20-2: Balancing a pole

Pole balancing is complex because the pole can tip in any direction around 360 degrees. By using an axle and two wheels, we can limit the directions a robot can tip to one plane. This means the object has only the option of tipping to the left or to the right (Figure 20-3). A Segway illustrates this concept nicely. Let's attempt to recreate this device with LEGO.

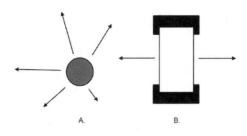

Figure 20-3: Balancing on two wheels

Building the Segoway

The balancing robot in this section is named Segoway, which is a mixture of the word LEGO with Segway (see Figure 20-4).

This project requires a single gyro sensor to measure rotational changes along an axis. You can obtain a gyro from HiTechnic.

> **NOTE:** In chapter 23 there is a project that uses two gyros. If you are interested in the other project, you might want to order two gyros at the same time to save money.

A gyroscopic sensor is one of the more costly additions to the NXT kit. However, according to press reports, it cost around $100 million to develop the Segway technology. By comparison, the $54.95 price for a gyro sensor is not too bad.

Figure 20-4: The Segoway

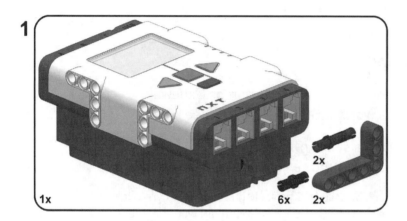

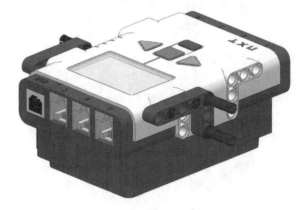

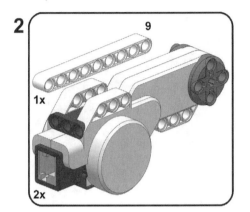

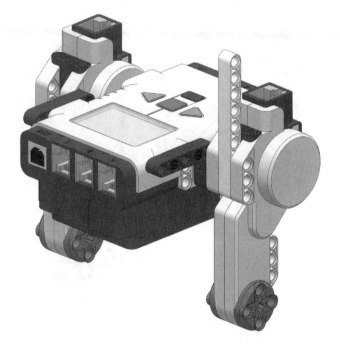

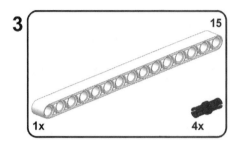

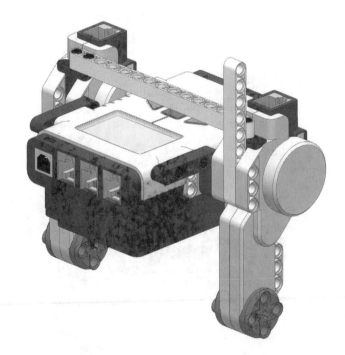

4

5

2x   2x

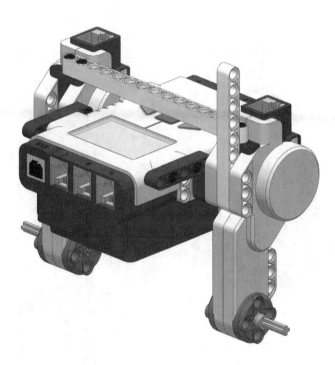

5

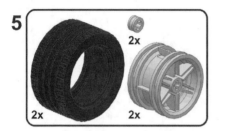

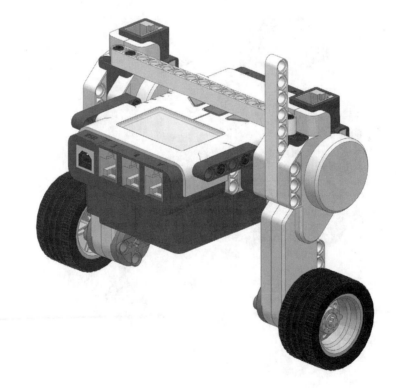

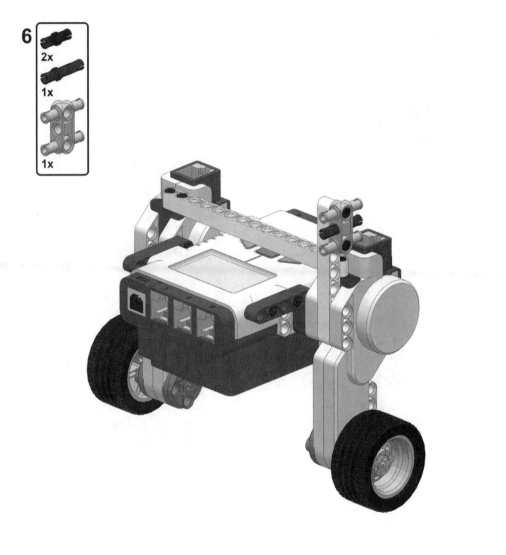

7

1x

1x

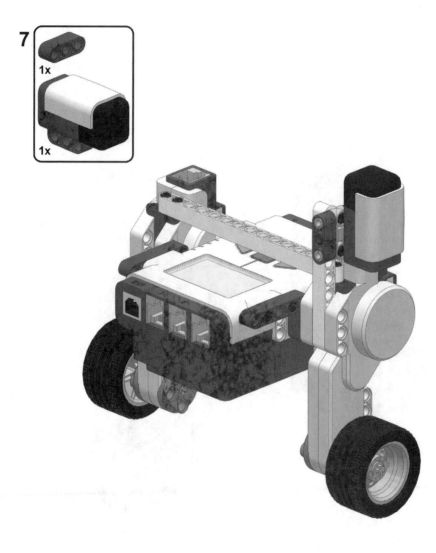

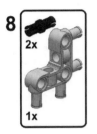

9

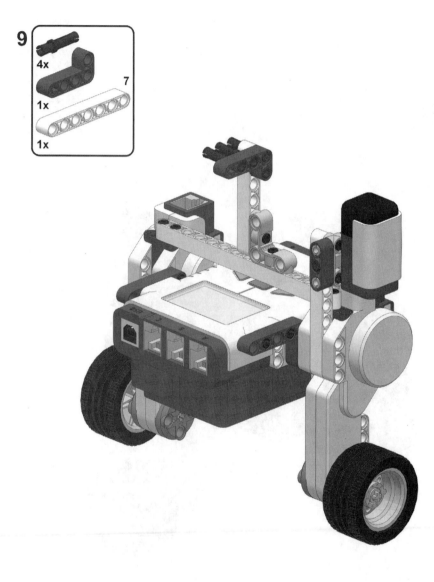

10

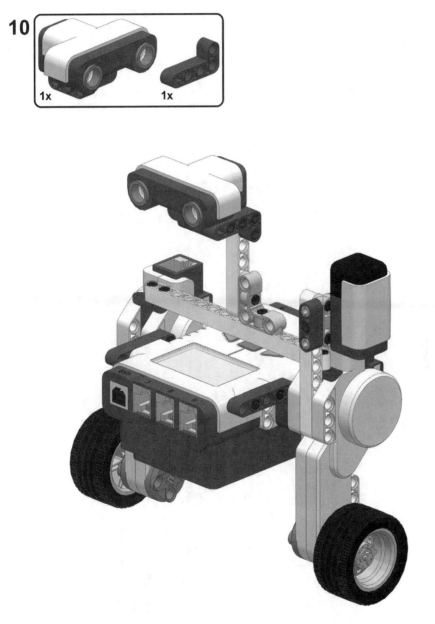

Connecting the cables

It is important to wrap the cables fairly tightly around the body of the robot so they are not loose. If they are loose, the cables will cause sympathetic vibrations and your robot will chatter as it moves.

1. Connect a short cable from the gyro to port 1.

2. Ports B and C connect to the motors using medium cables. Again, try to wrap the cables around the NXT to manage the excess length (see Figure 20-5). If you wrap the cables below the robot, they act as a nice cushion in case your robot tips over.

3. Finally, connect the ultrasonic sensor to port 4 using another medium cable.

Figure 20-5: Wrapping the cables around Segoway

Control Loop

A *control loop* is an engineering term for a machine that measures a value, reacts according to the value, and then repeats. The sensor has a target value that it continuously tries to attain, and any difference between the current value and the target value is called *error* (see Figure 20-6).

In this example, our control loop reads the gyro sensor value, determines which way it has tilted, and attempts to use the motors to compensate in the other direction. Sounds simple, right? Control loops in engineering are famous for sometimes overreacting and causing the machine to behave erratically. This project is no different. To understand how hard it is to achieve balance, try the following code and observe how it overreacts:

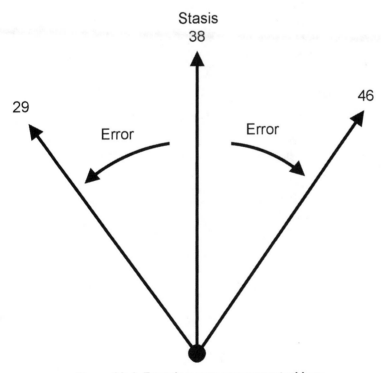

Figure 20-6: Error in a gyro sensor control loop

```java
import lejos.nxt.*;
import lejos.nxt.addon.GyroSensor;

public class SegowayBad {
  public static int speed = 0;
  public static int stasis = 0;

  public static void main (String[] args)
throws Exception {
    GyroSensor s = new GyroSensor(SensorPort.
S1);
    Motor.B.flt();
    Thread.sleep(1000); // give robot time to
remain still
    stasis = (int)s.getAngularVelocity();

    for(int i=0;i<5;i++) {
      Sound.beep();
      Thread.sleep(1000);
    }

    while (!Button.ESCAPE.isPressed()) {
      int aVelocity = (int)
s.getAngularVelocity();

      int diff = aVelocity - stasis;
      speed = Math.abs(diff) * 30;

      Motor.B.setSpeed(speed);
      Motor.C.setSpeed(speed);

      boolean forward = diff > 0;
      if (forward) {

        Motor.B.forward();
        Motor.C.forward();
      } else {
        Motor.B.backward();
        Motor.C.backward();
      }
      Thread.sleep(8); // 8 ms
    }

  Motor.B.flt();
  Motor.C.flt();
  }
}
```

To use this example, set down the robot so it is perfectly still and execute the program. After a few seconds it will start beeping. Stand it up and when the beeping stops the program will take over. The robot goes forward when the angle velocity is positive and backward when it is negative. I even went so far as to change the speed of the motor to reflect the amount it is tipping at.

This robot is very unstable and moves violently back and forth. Notice that the robot goes too far one way, and then goes too far the other way repeatedly? It doesn't learn that it is overshooting the mark each time and eventually it just falls over. We need to do something to control this behavior.

Programming Segoway

Above you saw a code snippet for a non-working inverted pendulum. So what do we have to do to make the balancing equation actually work?

There are four states which the Segoway code monitors:

- Gyro Speed – the angle velocity produced by the gyro sensor
- Gyro Angle – the angle the robot is tilting. Positive degrees when leaning forward, negative for backward
- Wheel Position – the position of the motor axle, in degrees. Read from the motor tachometer.
- Wheel Speed – the velocity of the motor, in degrees per second.

The Segoway algorithm runs in a control-loop, constantly updating these values and using them to determine the motor power for both motors. The equation looks like this:

```
Power    = Gyro Speed * Constant 1
         + Gyro Angle * Constant 2
         + Wheel Position * Constant 3
         + Wheel Speed * Constant 4
```

The result of this equation is that if the robot is only slightly out of balance, the equation will calculate a small number for power. After all, it will only need a slight nudge to re-balance. However, if several of the above factors are out of balance (it is tilted and it is tilting at a fast speed) then the power figure will be much larger.

The Segoway code is much more complicated than this. There are considerations for handling gyro readings, handling gyro drift, and moving the robot. Let's try using the Segoway class to control the robot.

```java
import lejos.nxt.*;
import lejos.nxt.addon.GyroSensor;
import lejos.robotics.navigation.*;

public class SegBumper {

  public static void main(String[] args)
throws Exception {
    NXTMotor left = new NXTMotor(MotorPort.B);
    NXTMotor right = new
NXTMotor(MotorPort.C);

    GyroSensor g = new GyroSensor(SensorPort.
S1);
    UltrasonicSensor us = new
UltrasonicSensor(SensorPort.S4);

    Segoway robot = new Segoway(left, right,
g, SegowayPilot.WHEEL_SIZE_NXT2);
    Thread.sleep(250000);
    robot.wheelDriver(-120, -120); // Move
forward
    while(!Button.ESCAPE.isPressed()) {
      if(us.getDistance() < 40) {
        robot.wheelDriver(120, 60); // Arc back
        Thread.sleep(2000);
        robot.wheelDriver(-120, -120); // Move
forward
      }
      Thread.sleep(350);
    }
  }
}
```

Unlike three- and four-wheeled robots, dynamically balancing robots must have fast code and a fast processor. Fortunately, leJOS and the NXT are more than up to the challenge. The balance control-loop in the Segoway class only takes 1.128 ms to run. At the end of every control loop, it sleeps for 8 ms. This means leJOS is about seven times faster than it needs to be. In fact, leJOS could simultaneously handle seven balance loops for seven Segway robots.

Using Segoway

To use Segoway, run the program and put down the robot so it is completely still. The program will take some readings from the gyro, then you will hear it start beeping. This is your signal to pick it up and place it in an upright position. When the beeping stops, Segoway will take over and maintain balance.

After a few seconds it will begin moving forward until it detects an object, at which time it will back up and move in a new direction.

If you're curious, try throwing some objects at the robot (try the other two rubber tires in the NXT kit). The Segoway should be able to maintain balance, even under stressful conditions.

> **NOTE:** You must have fully charged batteries. If your batteries are weak, the NXT will not be able to move the motors fast enough to maintain balance.

Just for fun!

The actual Segoway code listing is far too long to list in this book, and likely it would bore you. If you wish to view the code, it is available on Sourceforge at the following site:

http://lejos.svn.source-forge.net/viewvc/lejos/trunk/classes/

The Segoway code is located in lejos.robotics. navigation.

Is Taller Better?

The lejos.robotics classes are adaptable to a variety of robot designs. Likewise, the Segoway code is designed to work with almost any robot design—although there are limitations of course. You can easily design your own Segway-like robot and see how well it balances.

> **NOTE:** When building your own balancing robot, if you notice chattering when it is trying to balance, this is because a part on the robot is either not secure (such as motors that aren't reinforced very well to the main structure) or a part such as a long beam or cable is flexing and causing a sympathetic vibration.

Does the height of the robot improve balance (see Figure 20-8)? It seems intuitive that a taller robot would be easier for the NXT to balance because with humans, a long pole is easier to balance on your finger than a short pole (try balancing a pencil on your finger—it's seemingly impossible). So let's give it a try.

The tall robot balances in place fine, but you'll notice it isn't very stable. If you throw objects at it or make it move, it will fall down eventually. Why is that? If it's easier to balance a long pole, shouldn't a tall robot balance better?

Let's first examine why it's easier to balance a long pole on your finger. The key reason is a physics rule called the Moment of Inertia (MoI). The MoI has to do with how much force is required to cause

Just for fun!

It's surprising how well the Segoway code works with different robot designs. Try adapting the Carpet Rover robot with a gyroscope (see Figure 20-7). Using the same code you used above (minus an ultrasonic sensor), tip it on its two wheels and see how well it works. Surprised?

Figure 20-7: The Carpet Rover balancing with a gyro

an object to rotate. A longer and more massive object takes more force to overcome inertia, and hence it takes longer to rotate a long pole compared to a pencil. The distribution of mass is also important. A mass farther from axis of rotation requires more force to overcome inertia—a pole with a weight at the top requires more force. It's actually the same principle as a tightrope walker using a long heavy pole to help maintain balance.

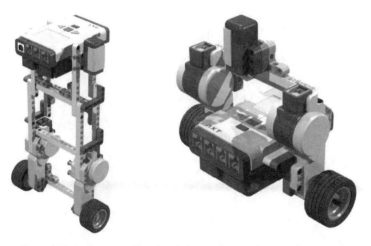

Figure 20-8: Does a tall robot balance better than a short one?

So why isn't a taller robot easier for the NXT to balance? There are two primary reasons.

First, the mass remains about the same, therefore the moment of inertia is not significantly changed. But center of gravity is higher, resulting in a more unstable robot.

Second, imagine if you upsized your robot so it is a mile high. But imagine your mile-high robot still has tiny NXT wheels. Once the robot starts tipping, the wheels simply can't rotate fast enough to keep up with the tilt. After all, the mile high robot will be moving very fast. You would need either super-fast motors to keep up with the tilt, or very large wheels (see Figure 20-9).

Figure 20-9: A mile-high robot

The motors would also need extra force to counteract the momentum once it starts falling. When the mass tips over it has fallen slightly, so in effect the motors need to lift the mass back up (and quickly), while moving in the same direction it is falling. So the motors need to be able to move quite a bit faster and push with more force to stay ahead of the tilt.

To put this to the test, try the tall robot with a freshly charged NXT (or you can even buy a set of 1.5 V alkaline batteries) which will supercharge the motors. You can also put on larger LEGO wheels. Now try throwing objects at the robot to see how well it copes.

SegowayPilot and Navigation

You have probably already noticed the SegowayPilot class in the same package as Segoway. This class is a typical pilot class and is therefore capable of calibrated moves and turns. This means SegowayPilot can interact with the classes that other pilots can, such as Navigator.

```java
import lejos.nxt.*;
import lejos.nxt.addon.GyroSensor;
import lejos.robotics.navigation.*;

public class SegwayPilotDemo {

  public static void main(String [] args)
throws InterruptedException {
    NXTMotor left = new NXTMotor(MotorPort.B);
    NXTMotor right = new
NXTMotor(MotorPort.C);

    GyroSensor g = new GyroSensor(SensorPort.
S1);

    SegowayPilot pilot = new
SegowayPilot(left, right, g, SegowayPilot.
WHEEL_SIZE_NXT2, 10.45F);
    pilot.setTravelSpeed(80);

    // Draw three squares
    for(int i=0;i<12;i++) {
      pilot.travel(50);
      pilot.rotate(90);
      Thread.sleep(2000);
    }
  }
}
```

The preceding class will trace out three squares on the floor. Try substituting a SegowayPilot for other projects in this book that use a pilot. It's fascinating to watch the Segoway navigate around the house!

CHAPTER 21

NXT
Communications

TOPICS IN THIS CHAPTER

- NXT Communications
- Segoway Remote Control
- USB Communications

In chapter 18 we found out how to easily write code to control the NXT brick. This code was almost identical to the regular leJOS NXJ code and it did not require any special communications commands. All of your program code ran exclusively on the PC.

In this chapter you will find out how to execute code on the NXT and on the PC simultaneously. The NXT code will handle time sensitive tasks that would be too slow to communicate from the PC, while the PC code will allow you to control the robot directly. To accomplish this, we will create a simple client-server protocol.

Communications can take place between the NXT brick and any other Bluetooth device. Not only can you communicate with other NXT bricks via your PC, but you can also communicate with smart phones, GPS units, handheld computers and other devices. We'll look at establishing communications between other devices in this book, but in this chapter we will focus on doing this with your PC.

NXT Communications Architecture

The goal of this chapter is to establish a Bluetooth connection between your PC and NXT brick. If you haven't already done so, pair your NXT with your computer (see chapter 18 for detailed instructions).

Once our Java code establishes a Bluetooth connection between a PC and NXT robot, there are a lot of things we can do. For example, we could make a user interface for a robotic arm. However, GUI projects take a lot of code, so to demonstrate communications we will make a simpler project. This project would be impossible to do with the simple PC Comms explored in the previous chapter.

We're going to control the Segoway robot from chapter 20 remotely using a PC. The Segoway class is not available in the PC Comms API. However, if you tried running the Segoway class on your PC (which would communicate through Bluetooth commands) it would run far too slowly to keep up with the real-time balancing requirements of the robot.

A typical architecture for communications is to have one application running on the NXT brick and another on the PC side (see Figure 21-1). To establish the connection, one device waits for communications to begin, while the other initiates the connection by contacting the other. The one that is waiting for a connection is called the *server*, while the one that connects is the *client*. Typically the client is the application that includes some sort of friendly user interface. We'll start with the server.

Smartphone Client

NXT Server

Desktop PC Client

Figure 21-1: Client-server architecture

Before you can use the server, build the Segoway robot from chapter 20 (see Figure 21-2). You'll need a gyro sensor for this project, so if you don't have one you can easily replace the Segoway robot with the Carpet Rover robot from chapter 3.

Figure 21-2: The Segoway robot ready to become a server

NXT Server

You might think that it would make more sense for the PC to be the server and the NXT to be the client. After all, a PC is more powerful than an NXT brick. In fact, the NXT makes a better server because it is desirable for the client to have a nice GUI. Also, the computer performing the end task is usually the server, and in this case, the NXT is performing the end task.

The NXT server has a few tasks it must perform:

- Wait for a Bluetooth connection
- Once connected, wait for commands
- When commands are received, move the Segoway
- When the connection closes, wait for another connection

Of course, this server is different from most other servers because it is a dynamically balancing robot. Let's enter the code.

```java
import java.io.*;
import lejos.nxt.*;
import lejos.nxt.addon.GyroSensor;
import lejos.nxt.comm.*;
import lejos.robotics.navigation.*;

public class SegowayServer implements
ButtonListener {

  public static final int LEFT = 1;
  public static final int RIGHT = 2;
  public static final int FWD = 3;
  public static final int BWD = 4;
  public static final int STOP = 5;
  public static final int QUIT = 99;

  static ArcRotateMoveController robot;

  public static void main(String[] args) {

    Button.ESCAPE.addButtonListener(new
SegowayServer());

    GyroSensor gyro = new
GyroSensor(SensorPort.S1);
    NXTMotor left = new NXTMotor(MotorPort.B);
    NXTMotor right = new
NXTMotor(MotorPort.C);
    robot = new SegowayPilot(left, right,
gyro, SegowayPilot.WHEEL_SIZE_NXT2, 16.5);

    while(true) {
      System.out.println("Server waiting.");
      NXTConnection conn = Bluetooth.
waitForConnection();
      DataInputStream in = conn.
openDataInputStream();
      System.out.println("Connected.");
      int reply = 0;
      do {
        try {
          reply = in.readInt();
          System.out.println("Command: " +
reply);
        } catch (IOException e) {
          e.printStackTrace();
        }
```

```
      performMove(reply);

    } while(reply != QUIT);

    try {
      in.close();
      Thread.sleep(100);
      conn.close();
    } catch (IOException e) {
      e.printStackTrace();
    } catch (InterruptedException e) {
      e.printStackTrace();
    }
  }
}
private static void performMove(int cmd) {
  switch (cmd) {
      case FWD:
        robot.forward();
        break;
      case BWD:
        robot.backward();
        break;
      case LEFT:
        robot.rotate(Double.POSITIVE_
INFINITY);
        break;
      case RIGHT:
        robot.rotate(Double.NEGATIVE_
INFINITY);
        break;
      case STOP:
        robot.stop();
        break;

  }
}

public void buttonPressed(Button arg0) {}

public void buttonReleased(Button arg0) {
  System.exit(0);
}
}
```

NOTE: If you don't have a gyro sensor, you can use the Carpet Rover instead and replace the SegowayPilot line with the following code:

```
robot = new DifferentialPilot(diam, width,
Motor.B, Motor.C, false);
```

Typically with Java communications, we use the stream classes found in the java.io package. As you can see above, once the connection is established, it retrieves a DataInputStream from the NXTConnection object using the open-DataInputStream() method.

You can try running this code even though the PC client is not programmed yet. The Segoway robot will prompt you to stand it up when it is ready, and then it will happily balance while it waits for a client to connect.

PC Client

Now it's time to switch gears and create a client to connect to the Bluetooth server. Here is where Eclipse comes in handy. Eclipse makes it very easy to switch back and forth between coding an NXJ app and coding a PC app. You can have the client class open in one tab next to the server class, and when you press the green run button, it will perform the appropriate action (either uploading the class to the NXT or running the program on your PC). If none of this makes sense right now, don't worry, it will when you start using it.

NOTE: If you are using command line, don't forget to add pc-comm.jar to your classpath. The jar file is located in the lib\pc in your leJOS install directory.

Normally a client has a graphical user interface (GUI) to easily allow you to click and control things. Our client will have a very simple window which accepts key-presses from your keyboard, similar to the example in chapter 18.

```
import lejos.pc.comm.NXTConnector;
import java.awt.event.*;
import java.awt.*;
import java.io.*;

public class SegwayClient extends Frame
implements KeyListener{

  // Protocol values:
  public static int ROT_LEFT = 1;
  public static int ROT_RIGHT = 2;
  public static int FWD = 3;
  public static int BWD = 4;
  public static int STOP = 5;
  public static int PROG_QUIT = 99;

  // Keyboard control values:
  final int FORWARD = 87, // W = forward
  BACKWARD = 83, // S = backward
  LEFT = 65, // A = left turn
  RIGHT = 68, // D = right
  QUIT = 81; // Q = quit

  boolean steerreleased = true;
  boolean drivereleased = true;

  NXTConnector conn;
  DataOutputStream dos;

  public SegwayClient(String title) {
    super(title);

    conn = new NXTConnector();
    boolean connected = conn.connectTo("btspp://");

    if (!connected) {
      System.err.println("Failed to connect to
any NXT");
      System.exit(1);
    }

    dos = new DataOutputStream(conn.
getOutputStream());

    this.setBounds(0, 0, 300, 50);
    this.addKeyListener(this);
    this.setVisible(true);
  }

  public void keyPressed(KeyEvent e) {
    switch(e.getKeyCode()) {
```

```
      case FORWARD:
        if(drivereleased==true) sendCommand(FWD);
        steerreleased = false;
        break;
      case BACKWARD:
        if(drivereleased==true) sendCommand(BWD);
        steerreleased = false;
        break;
      case LEFT:
        if(steerreleased==true) sendCommand(ROT_LEFT);
        steerreleased = false;
        break;
      case RIGHT:
        if(steerreleased==true)
    sendCommand(ROT_RIGHT);
        steerreleased = false;
        break;
      }
    }

  public void keyReleased(KeyEvent e) {
    switch(e.getKeyCode()) {
    case FORWARD:
    case BACKWARD:
    case LEFT:
    case RIGHT:
      sendCommand(STOP);
      steerreleased = true;
      break;
    case QUIT:
      sendCommand(PROG_QUIT);
      System.exit(0);
    }
  }

  public void keyTyped(KeyEvent e) {}

  private void sendCommand(int cmd) {
    try {
      dos.writeInt(cmd);
      dos.flush();
    } catch (IOException e2) {
      e2.printStackTrace();
    }
  }

  public static void main(String[] args) {
    new SegwayClient("Enter commands");
  }
}
```

Notice the following line above:

```
boolean connected = conn.
connectTo("btspp://");
```

Using btspp indicates we want to connect using Bluetooth. Normally the Bluetooth address of the device you want to connect with follows, but in this case if we leave it blank it will try to connect with any Bluetooth NXT units it finds in the vicinity.

NOTE: Assuming your NXT code is running and waiting for a connection, when you run the PC client code it will automatically connect. If it doesn't work the first time, try it again—sometimes Bluetooth connections fail. Once the GUI window pops up, you can control the robot using the W-A-S-D keys, or press Q to quit.

The java.io package contains numerous streams used in Java communications. These include the basic InputStream and OutputStream classes, which can send only bytes. To send larger data types, such as integer and double values, use the DataInputStream and DataOutputStream classes.

USB Communications

You can easily create a connection using USB, although obviously this is not suitable for NXT robots that are mobile. But if you have a stationary project, such as a robotic arm, the USB cable has even faster communications than Bluetooth.

You will need to alter the code on the server and client. On the server, replace the Bluetooth.waitForConnection() code with USB.waitForConnection(). On the client, use the following line of code:

```
boolean connected = conn.connectTo("usb://");
```

CHAPTER 22

Android

TOPICS IN THIS CHAPTER

- Google Android Platform
- Android Eclipse plugin
- Programming a leJOS Android app

Asmartphone with Bluetooth makes a good remote control for guiding robots. The unique properties of a smartphone make it ideal for robotics projects. They are small, portable, fit in your hand and can be attached to your NXT robot easily. As of this writing, the most ubiquitous smartphone operating system is Android by Google (see Figure 22-1).

An Android smartphone has many of the same benefits as your PC. The color display is ideal if you want to view data in real-time without chasing after your NXT robot to look at the LCD. You get much more memory than the NXT, allowing you to store large data files that your NXT collects. You also have access to more programmable input keys than the NXT and the touch-screen has interesting possibilities for robotic control. Finally, you get access to all kinds of features via the Android Java API, such as GPS (if your phone has it), Internet, SMS messaging, multimedia, and more.

Figure 22-1: An assortment of Android devices

NOTE: As of this writing, leJOS Android support is in an early stage of development. Check with our online tutorial for the latest Android instructions.

http://lejos.sourceforge.net/nxt/nxj/tutorial/

Installing the Android SDK

Before you can begin developing software for your Android device, you need to install the Android SDK. The SDK is available for Windows, Macintosh, and Linux. The Windows version includes an installation program.

Macintosh and Linux users should follow the instructions here:

Mac

http://developer.android.com/sdk/installing.html

1. Download the version for your platform here:

http://developer.android.com/sdk/

2. Install the Android SDK. Note: If the Windows installer does not detect your Java SDK, hit the Back button and click Next again.

3. When installation is complete it will ask to run the SDK and AVD Manager. Ensure the box is checked and click Finish.

4. The manager recommends packages to install (see Figure 22-2). These can take a long time to download. If you know which version of the Android OS you own, I recommend deselecting the other versions by selecting them in the list and clicking Reject. When done, click Install.

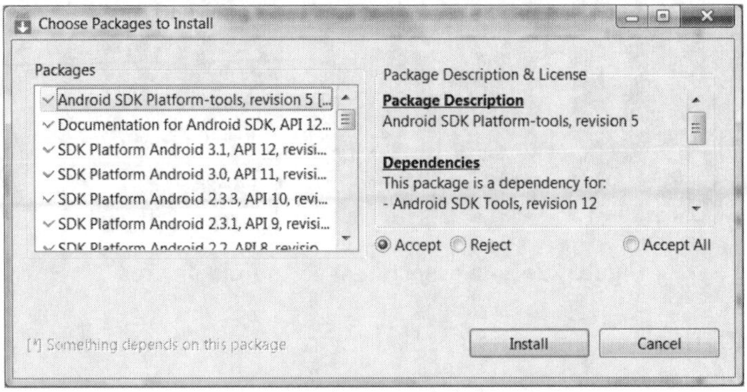

Figure 22-2: Choosing packages to install

Installing Eclipse Android Plugin

Now that you have the Android Java SDK installed, we need to install an Eclipse plug-in called Android Development Tools (ADT).

1. Run Eclipse. Select Help > Install New Software...
2. Click Add in the upper right corner. For name enter ADT Plug-in (see Figure 22-3). For location enter:

   https://dl-ssl.google.com/android/eclipse/

Figure 22-3: Adding the ADT plug-in repository

3. After a moment you should see Developer Tools. Select the checkbox next to this and click Next.
4. In the next window, you'll see a list of the tools it will download. Click Next.
5. Read and accept the license agreements, then click Finish.
6. When the installation completes, click Restart Now.
7. Now we need to configure the plug-in. Select Window > Preferences > Android.
8. Browse to the location you installed the SDK (see Figure 22-4). Click Apply after browsing to confirm your choice and you should see the version of the SDK you downloaded earlier.

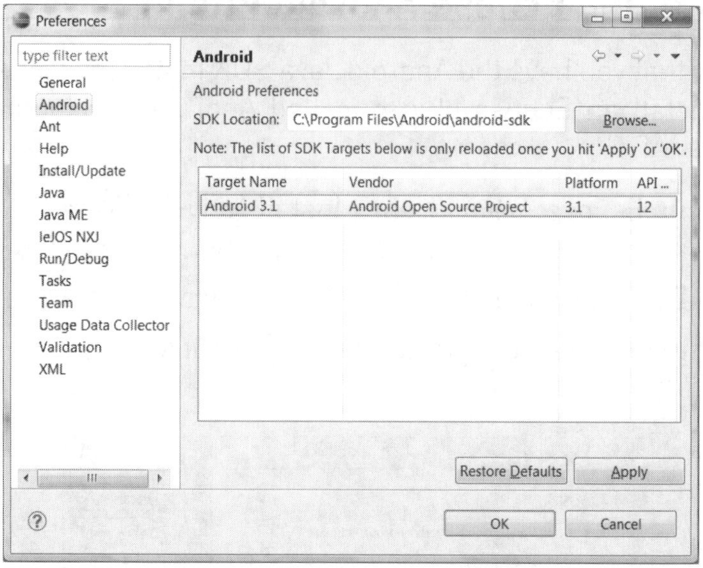

Figure 22-4: Specifying the SDK location

You are now all setup to begin developing.

Running leJOS Android Code

Android applications end with .apk. The leJOS project has a sample application programmed for Android called leJOSDroid.apk, which contains multiple samples. These applications are included in your leJOS install directory in the projects folder. If you want to change this code, you will need to import the project into Eclipse.

1. In Eclipse, select File > Import... Expand the General folder and select Existing Projects into Workspace (see Figure 22-5). Click Next.

2. Browse to the leJOS install directory. Select projects\android\leJOS-Droid. Make sure the leJOS-Droid folder is selected (see Figure 22-6). Click Finish.

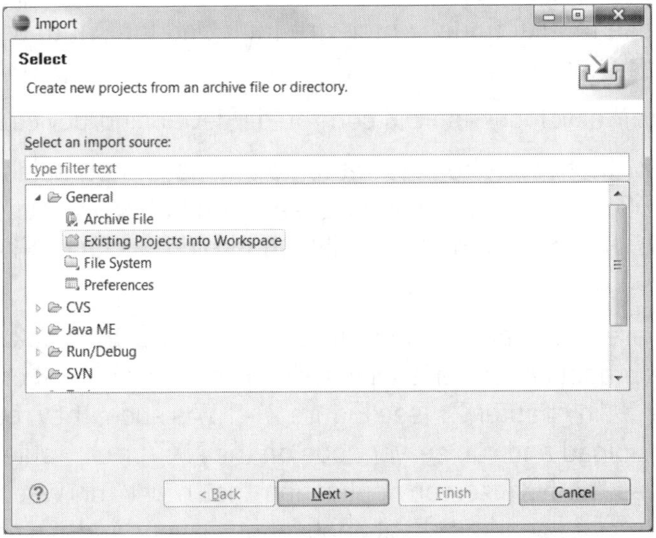

Figure 22-5: Importing the leJOS-Droid project

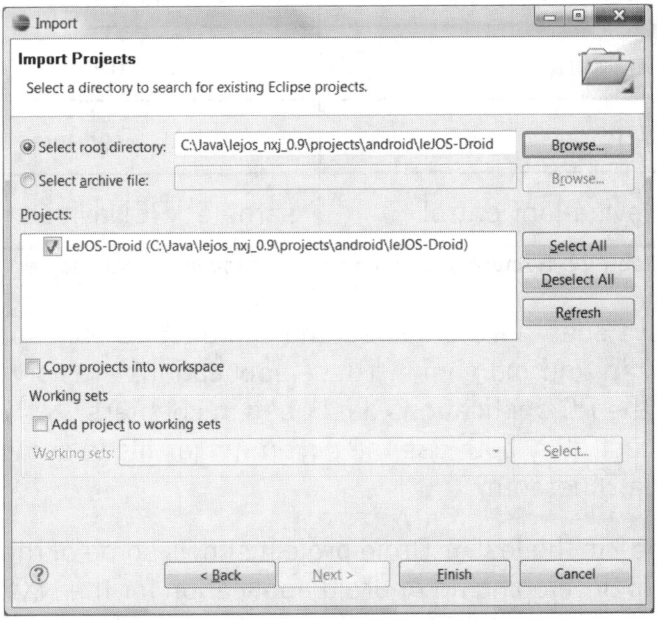

Figure 22-6: Importing the leJOS-Droid project

3. Before you can upload this code to your device, you must setup your device. The following website contains these instructions, which are included for Windows, Mac OSX and Linux:

   http://developer.android.com/guide/developing/device.html

4. Now that your phone is setup, to upload, simply select LeJOSDroid.java and click the green Run button in Eclipse. This will launch the ADP utility (Android Developer Phone).

There are three applications in the file leJOSDroid.apk. Two of these applications are typical client-server applications as explained in chapter 21 (see Figure 21-1). As such, they require you to upload and run server code on the NXT brick. Table 22-1 specifies the NXT server application that must run when you run the corresponding client application on your Android phone.

Android Client App	NXT Server App
TachoCount	none (main menu)
BTSend	samples\BTReceive
RCNavigationControl	samples\RCNavigator

Table 22-1: Android client app and corresponding NXT server

We have seen how to upload and run a leJOS NXJ application to an Android phone. These applications are very similar to the PC applications described in chapters 18, 19 and 21. In fact, they both use the pccomms.jar file to communicate via Bluetooth.

The code in the leJOS-Droid project shows some of the specifics of developing an Android application for the NXT. The specifics of Android programming are beyond the scope of this book, but if you want some tips, visit the tutorial section on leJOS as mentioned above.

CHAPTER 23

Ballbot

TOPICS IN THIS CHAPTER

- Ballbot
- LEGO NXT Orbot
- Movement

In chapter 20, a two-wheeled robot used a gyro to achieve dynamic stability along a single axis of balance. This chapter attempts dynamic stability using a robot that balances on a ball. As you probably suspect, this is more complicated and requires data from an additional axis, meaning it requires an additional gyro sensor.

Ballbot

A ballbot is a type of robot that consists of a robot balancing on a sphere. The first ballbot teetered across a research lab in 2006. It was created by Ralph Hollis and his team from Carnegie Mellon University. Since then, there have been a rapid number of improvements on the design, such as a robot named Rezero from a university in Zurich, Switzerland (see Figure 23-1).

Figure 23-1: Nimble Rezero balancing

The method of dynamic stability for a ballbot is similar to that of the Segoway in the previous chapter. The major difference is that a ballbot is balanced on the point of a sphere, so it is not just unstable along one axis like Segoway, but rather it can tip in any direction—360 degrees around (see Figure 20-3). So how do we keep it from falling over?

The solution comes from the method of control of the ball. There are two motors for the control mechanism. These two motors rotate the ball. One motor rotates the ball along the x-axis, and the other rotates it around the y-axis (see Figure 23-2). The mechanism is similar in design to a mechanical computer mouse (see Figure 23-3), except that motors are attached to the rollers instead of sensors.

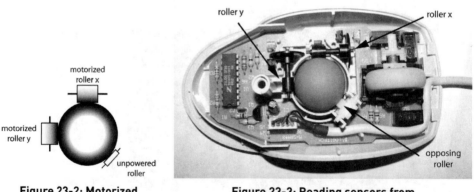

Figure 23-2: Motorized rollers rotating the sphere

Figure 23-3: Reading sensors from a mechanical mouse

By rotating the motors at different speeds, the ball can rotate in any direction, even though there are only two control axes. Therefore, one axis is controlled by a dynamic stability algorithm running in one thread, and the other axis is controlled by the same algorithm running in parallel in another thread (see Figure 23-4).

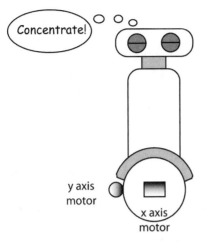

Figure 23-4: A ballbot running two threads at once

Orbot

There are several ballbot designs, many using three powered rollers in a triangle configuration to drive the ball. The LEGO NXT ballbot in this section, named Orbot, uses the classic mechanical mouse design of two perpendicular rollers with an unpowered roller on the opposite side to hold the ball in place (see Figure 23-5).

Just for fun!

An accelerometer is not capable of performing dynamic stability. You might be tempted to use one for this project, since LEGO accelerometers provide data in three axes. However, the mechanism of the accelerometer is not the same as a gyro. When the robot is moving and tilting at the same time, the acceleration readings will cancel each other out, resulting in the ballbot being fooled about the current state.

Figure 23-5: Orbot balancing on a plastic LEGO ball

Part selection is important for this project. Orbot uses only parts found in the NXT kit. If you want to confine your creation to 100% LEGO parts, the NXT 1.0 kit contains a ball. Unfortunately, owners of the 1.0 kit require tires from the 2.0 kit (see Table 23-1). If you do not have access to these parts, you can order them online from LEGO. You do not need to order the plastic hub for the tire because it is the same size in both the NXT 1.0 and 2.0 kits. Only the rubber tires are of different sizes.

You can also substitute the LEGO ball with an alternate ball of similar diameter. High friction and low inertia are essential qualities. Even though the LEGO ball has low friction and low inertia, it still works well as long as you can obtain a small bottle of rubber cement—more on that later.

The diameter of the LEGO ball is exactly 5.2 cm (2.05 inches). This is close to the diameter of a billiard ball, which is 5.72 cm (2.25 inches). If you remove the two black plastic gears on the opposing roller and replace them with two bushes, the billiard ball fits nicely in the chassis. The NXT tires are flexible, and the opposing roller has some give, so the ball does not have to be an exact match.

An even better substitute is a ball with more friction, such as a rubber ball. A racquetball is exactly 5.72 cm (2.25 inches) and available at sports stores. It will be able to grip the floor surface better than a hard plastic ball. You can find an assortment of balls in the toy department of a retail store. If the ball is smaller than 5.2 cm diameter, try adding larger gears to the roller in order to compensate for the smaller ball.

Name	LEGO Part No.	Image
Technic Ball 52mm	41250	
Tire 43.2 x 22 ZR	44309	

Table 23-1: Acquiring the unique LEGO Ballbot parts

Just for fun!

Why does the LEGO NXT only have a gyro with one axis and not three? It comes down to speed. To return multiple values, a sensor must use the I²C bus. However, normally I²C ports are not as quick as analog sensors. LEGO NXT-G and other firmware replacements do not support high-speed I²C, therefore the sensor makers did not bother producing a high-speed I²C gyro sensor. Note: The leJOS JVM implements high-speed I²C ports, which return data fast enough for use in time-critical dynamic stability applications like the ballbot.

1

9
15
7

1x
1x
1x
2x

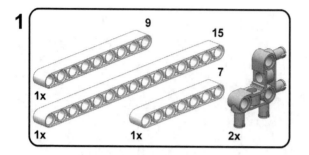

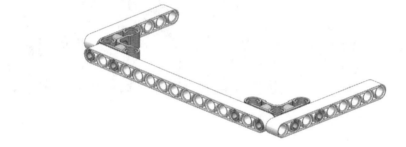

2

7

1x
1x
4x

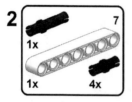

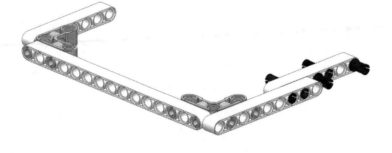

3

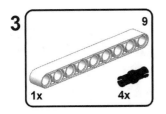

9

1x   4x

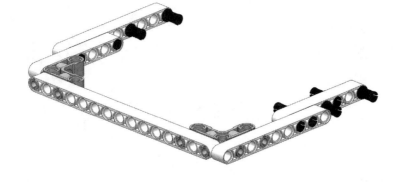

4

7

2x

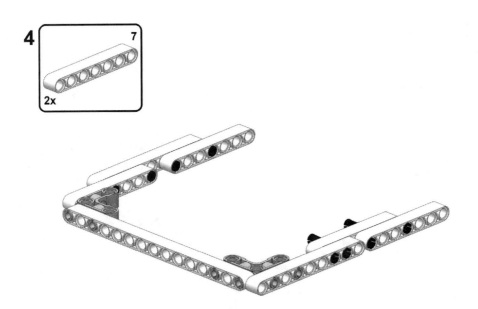

5

2x

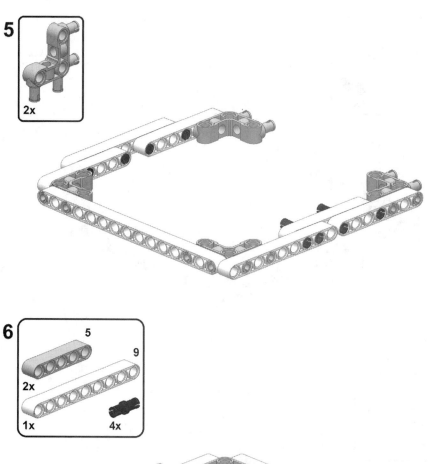

6

5

9

2x

1x   4x

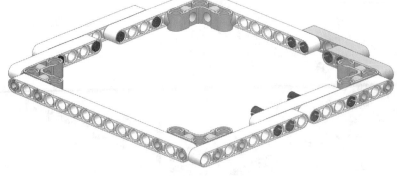

7

4x

2x

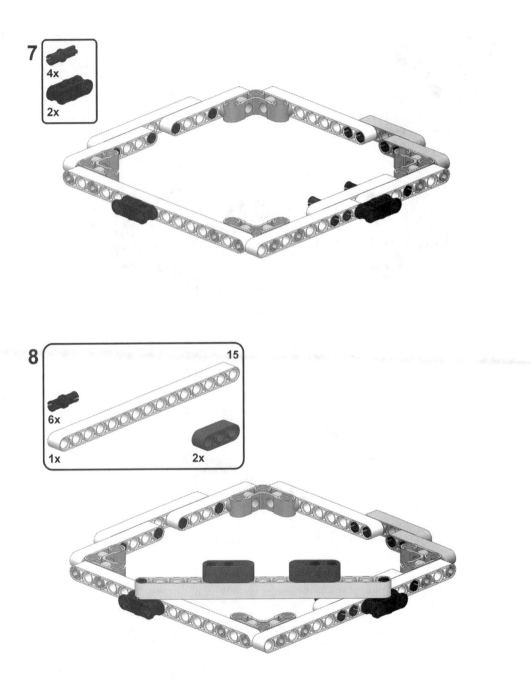

8   **15**

6x

1x   2x

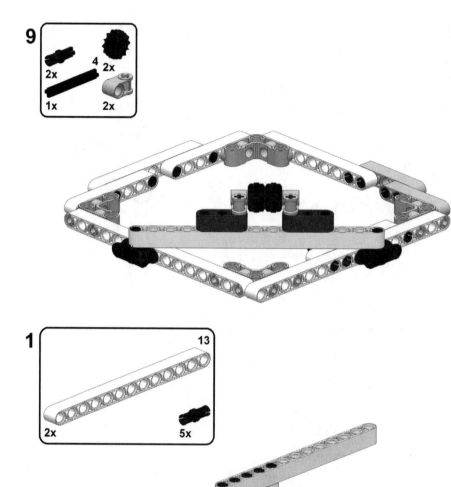

2

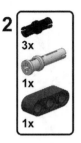

3x
1x
1x

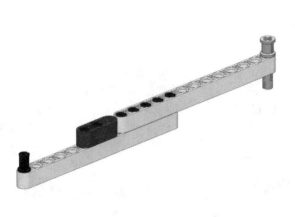

10

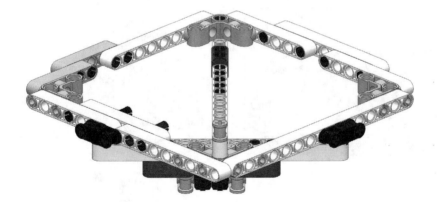

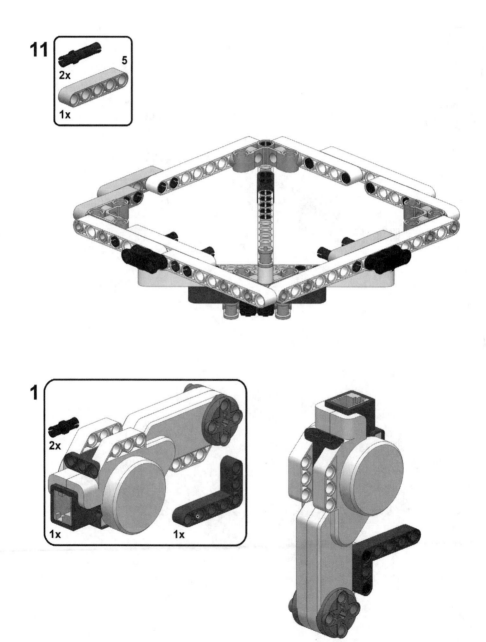

2

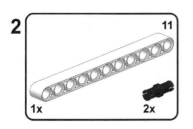

1x 2x

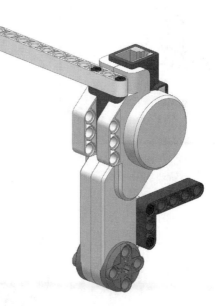

3

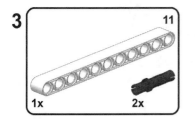

1x 2x

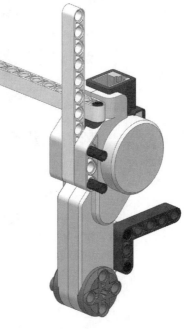

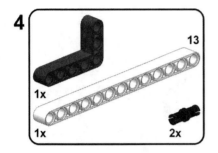

5

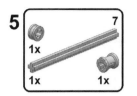

7

1x
1x   1x

6

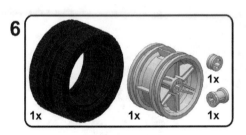

1x
1x   1x   1x

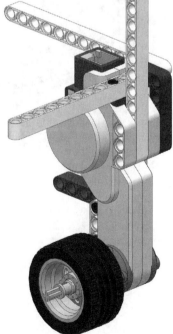

12

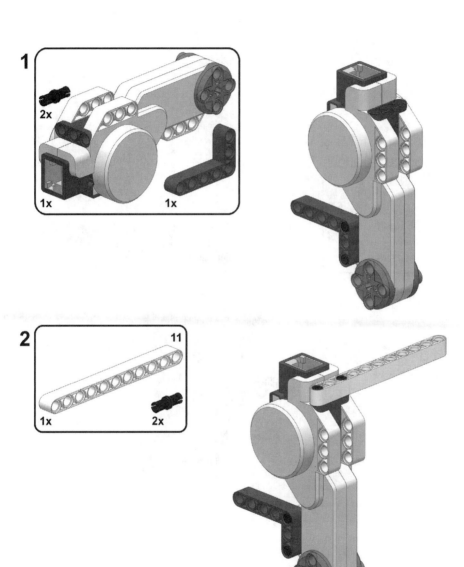

3

11

1x 2x

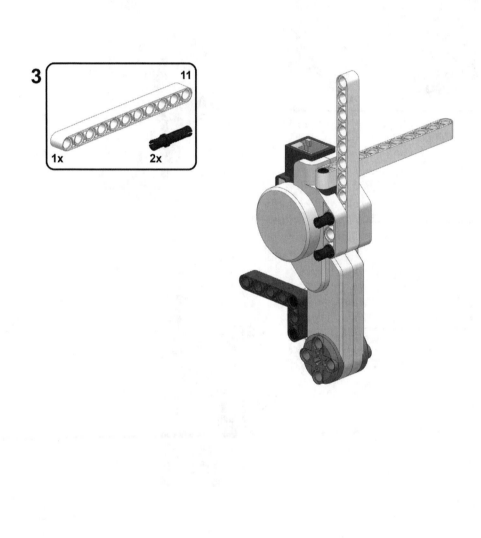

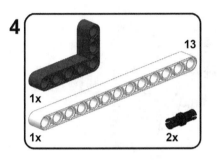

5

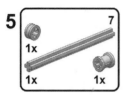

7

1x

1x

1x

6

1x

1x

1x

1x

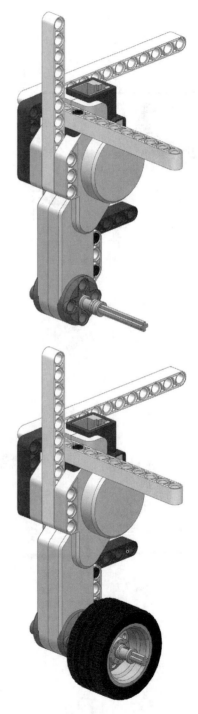

13

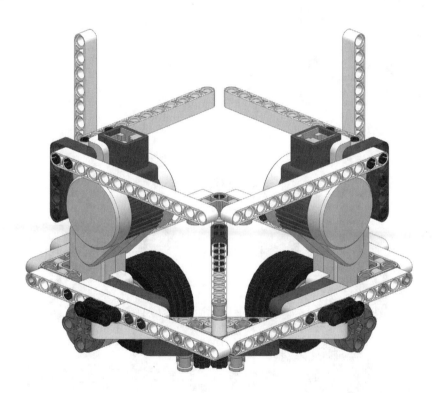

14

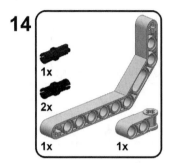

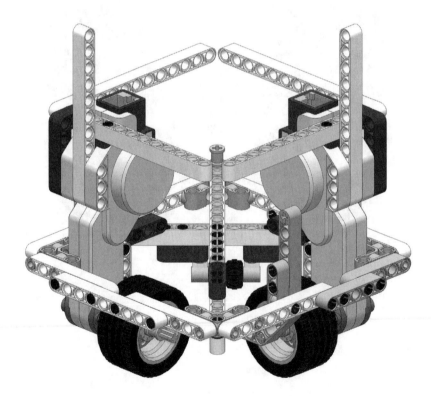

15

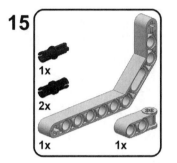

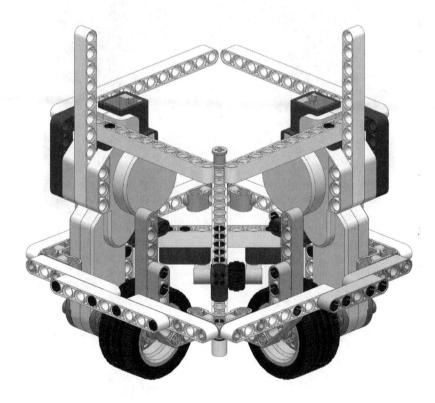

16

1x

2x

1x

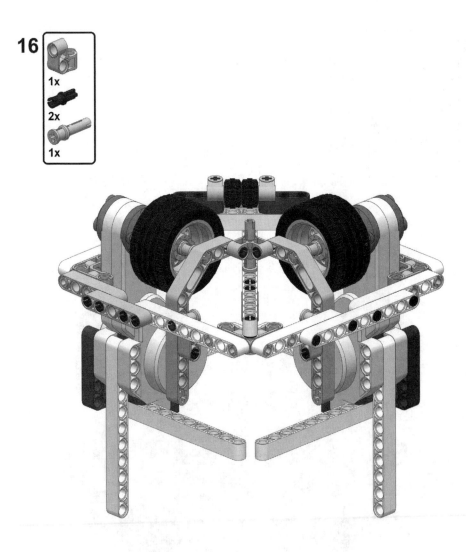

17

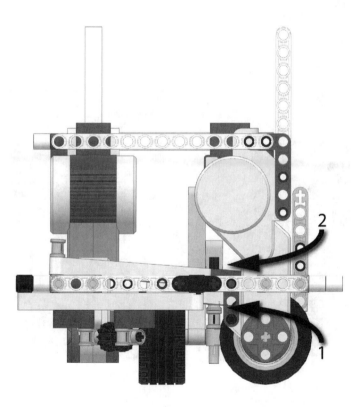

18

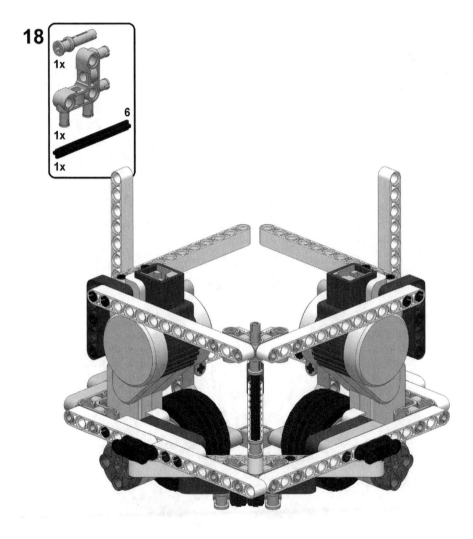

19

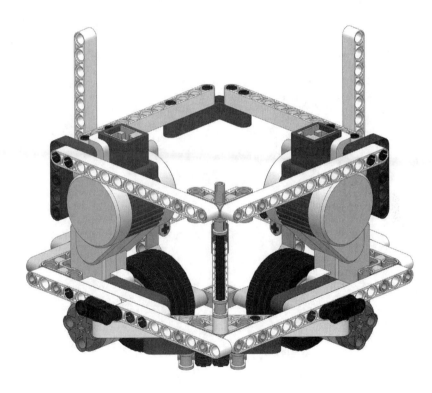

20

4x

2x

2x

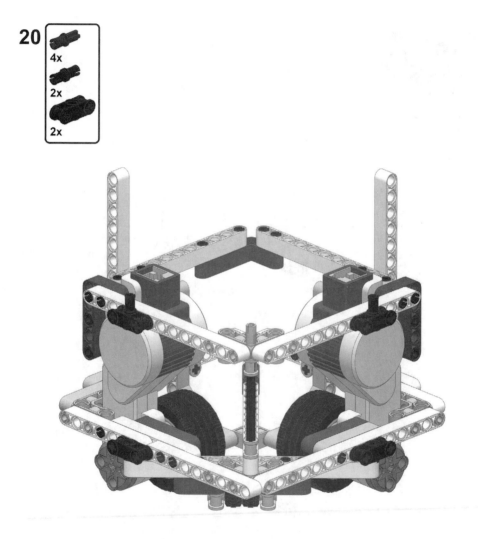

21

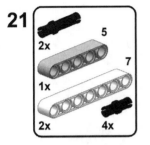

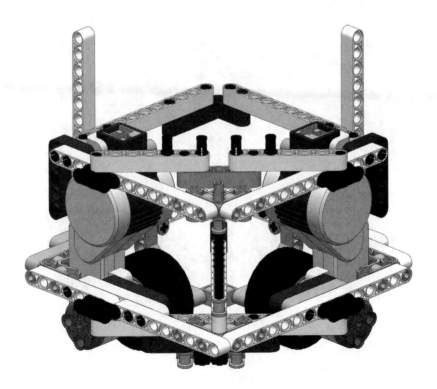

22

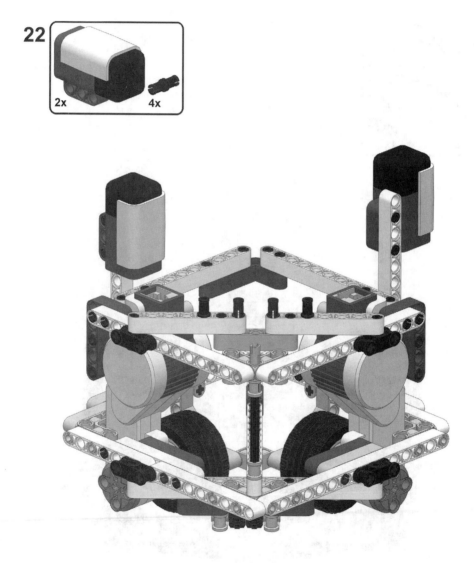

23

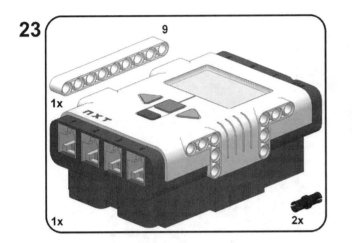

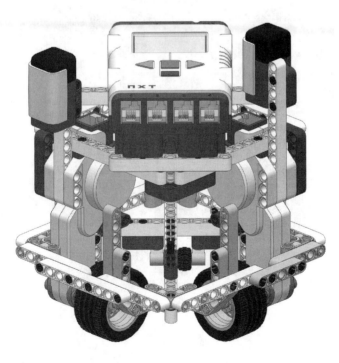

24

1x

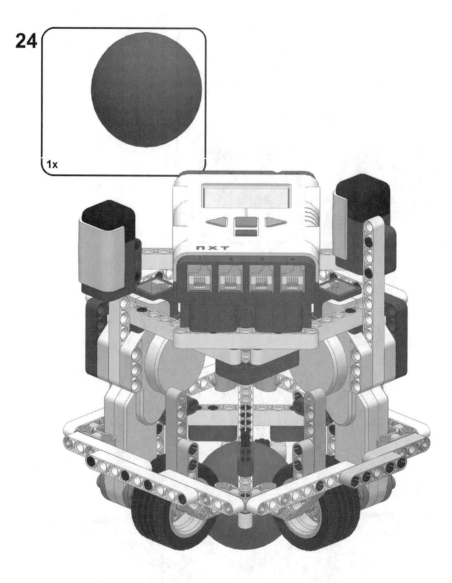

The ball mechanism is based on the NXT Ballbot by Tomás Arribas from the University of Alcala de Henares in Madrid, Spain.

Programming Orbot

The Ballbot class in the lejos.robotics.navigation package contains the balancing algorithm. This algorithm is almost identical to the Segoway algorithm, except two threads are running simultaneously instead of one. If you wish to browse the source code of the Ballbot class, see chapter 20 for instructions on viewing code on Sourceforge.

```
import lejos.nxt.*;
import lejos.nxt.addon.GyroSensor;
import lejos.robotics.navigation.*;

public class Orbot {
  public static void main(String [] args)
throws Exception {

    float tireCompression = 0.3F;

    NXTMotor xMotor = new
NXTMotor(MotorPort.B);
    GyroSensor gx = new GyroSensor(SensorPort.
S2);

    NXTMotor yMotor = new
NXTMotor(MotorPort.C);
    GyroSensor gy = new GyroSensor(SensorPort.
S4);

    Ballbot bb = new Ballbot(xMotor, gx,
yMotor, gy, MoveController.WHEEL_SIZE_NXT2-
tireCompression);

    Button.ESCAPE.waitForPressAndRelease();
    System.exit(0);
  }
}
```

Results

As previously mentioned, the plastic LEGO ball has low friction. This means it will not produce good traction on slippery hard floors. Soft linoleum floors are the premier surface for this type of robot, followed by short carpet, hardwood, and tile being the worst.

When you first run the Ballbot code, you will need to set the robot down so it is perfectly still while both gyros calibrate. Once this is done, an audible countdown will begin. You have eight seconds to balance the robot on the ball. Try holding it lightly in your fingers so you are barely keeping it stable. When the countdown ends, release the robot and the dynamic stability algorithm will take over.

Problem	Solution
Robot starts normal but jitter becomes worse and it tips over	Cables are not wound tightly enough to the body, causing reverberations
Robot starts normal but eventually starts wandering and tips over.	Batteries are not fully charged.
Robot stays up for about 5 seconds then tips over	The tires have grip at the start, but they polish the surfaces and start slipping. Increase tire friction with rubber cement.

Table 23-2: Identifying problems with your Ballbot

Ballbot is not as stable as Segoway. The reason is because the two motors are acting against each other. If you look at Orbot from above, it is symmetrical along the long cross brace (see Figure 23-5). However, the weight distribution of the model is not totally symmetrical all around.

When the x-axis motor corrects tilt, it causes the weight distribution to change slightly along the y-axis. In turn the y-axis motor corrects for this, which causes another weight distribution to change. This goes back and forth, rarely coming to a nice equilibrium the way Segoway does. A more symmetrical (with regard to weight) robot would perform better. Most ballbots from research institutions are symmetrical all around (see Figure 23-1).

Orbot can be a little temperamental. When using 100% un-modified LEGO parts, it works pretty well at first, but after a while the friction of your tires and ball will decrease. This is because the rapid movements of the LEGO wheels tend to polish the ball and tires, making them even smoother. This causes the ball to slip more against the tire, and the robot will tip over more easily.

Orbot works well with new LEGO tires that are relatively fresh out of the NXT kit and have not been used in dirty environments. However, they will eventually become polished and lose friction. If your robot tips over after only a few seconds, the tires are probably not gripping the ball. I found the most effective solution to this problem is to apply a layer of rubber cement to the treads (see Figure 23-6). You can obtain a small bottle of rubber cement from any office supply store for a few dollars.

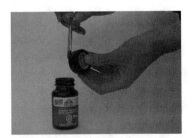

Figure 23-6: Applying rubber cement to the tires

Just for fun!

Now that you know how to make Orbot move, try using some of the remote control code from earlier in the book. Imagine controlling Orbot with an Android phone!

Movement

You can also cause Orbot to travel using the impulseMove() method in the Ballbot class. This method accepts a speed value ranging from -200 to 200 for each axis. Realistically you will need to use values around -20 to 20, otherwise Orbot destabilizes. By feeding two parameters, the Ballbot class will combine movement along both axes to create diagonal movement, making Orbot a true omnidirectional robot. Try adding the following code to the sample above to cause Orbot to move. The line should appear just below the constructor.

```
bb.impulseMove(20, 0);
```

Orbot does not move consistently like a regular robot because it needs to use the motors to maintain balance, but it will wander along the axis you select.

CHAPTER 24

Visionary Robots

TOPICS IN THIS CHAPTER

- Image Tracking
- Predator Turret

Vision is one of the most exciting and complex robotic tasks. There are two basic uses for vision: as telepresence for yourself, and as a sensor for your robot.

Most robots do not have vision in the sense that humans do. Instead, they cheat by scanning with a range finder, which reports back the distance to objects. Why do most robots measure distance instead of using cameras to see the whole picture? Because humans can pick out data from the images in our brain, whereas computers still have problems making sense of complex imagery.

This chapter will use a video camera to make a hunting robot. If you don't have a camera you can still build the robot and use an ultrasonic sensor instead.

A Video Camera

The LEGO NXT brick has a video camera designed just for robotics. The camera is small and light, and it connects directly to the NXT brick via a sensor port, so your robot won't be tethered to your computer via Bluetooth or USB cable (see Figure 24-1). The camera is much more complex than a webcam. It contains a real-time image processing engine for tracking objects based on color criteria, and it can also track lines. It currently does not detect shapes, though a firmware upgrade could allow this in the future. This camera, NXTCam, is designed and marketed by Mindsensors.com.

NXTCam captures at 30 frames per second and a resolution of 88 x 144 pixels, which is very low, but higher resolutions would require more processing power. No external battery pack is required because it is powered by the I^2C port of the NXT brick.

A mini-USB port allows the camera to connect to your computer. The camera does not come with the required mini-USB cable, which is sold separately by Mindsensors. If you have a digital camera or phone you probably already have one (it requires a five wire cable—a four wire version will not work). The camera is a little on the pricey side—$149.75 at the time of this writing.

Figure 24-1: Mindsensors NXTCam V3

The NXTCam is not a webcam. It cannot deliver real time video to your PC. Instead, it merely tracks objects and sends object data back to the NXT or your PC. It has an optimal focus of between two and four feet. You can adjust the lens manually to optimize focus for other ranges. The lens blocks infrared (IR) light from entering the lens. If you want to see IR light, you can order a special infrared permitting lens from Mindsensors (see Figure 24-2).

Figure 24-2: Infrared permitting lens

The camera needs some configuration before it can detect colored objects. In the next few sections we will learn how to install the USB driver, how to connect the camera to a computer to see what it sees, and how to change the camera settings so we can use it in a project.

Installing USB Drivers

Before you can use the NXTCam with your computer, you will need to install the proper USB drivers. As of this writing, these are available for Windows Vista, Windows XP, and Mac OS X. You can download the appropriate driver from the Mindsensors website.

Mac

www.mindsensors.com/NXTCam_Driver_Installation.htm

Windows Vista

1. Download the drivers and extract in a folder, such as: C:\Users\Admin\nxtcam_v1p1

2. Connect the camera to your USB port and wait for the Found New Hardware window to appear. In this window, select Locate and Install driver software (and give permission to continue).

3. When it asks you to insert a disc, select *I don't have disc, show me more options.*

4. Select *Browse my computer...* and select the folder C:\Users\Admin\nxtcam_v1p1\nxtcam

5. Vista may popup a window saying *Cannot verify the publisher of this driver software...* Choose *Install this driver software anyway.*

6. You aren't done yet! In a few seconds another window will pop up. Repeat from step 3 for the second driver. When this completes it will list a communication port, such as Com9.

Windows XP

1. Download the USB Driver ZIP file to your disk and un-zip it into a folder, such as c:\nxtcam_v1p1

2. Connect the NXTCam to a USB Port. The New Hardware Wizard will pop up.

3. In the wizard, select *Install from a list or specific location.*

4. Next, uncheck *Search removable media* and check *Include this location to search.*

5. Specify the path with an 'nxtcam' subfolder, such as c:\nxtcam_v1p1\nxtcam

6. In the next step, a dialog may inform you that the driver has not passed the Windows Logo test for compatibility. Select *Continue Anyway.*

7. In the next step, if a dialog box appears asking for the path of sys files, provide the path you gave in prior step followed by \i386. (or directory for your processor) e.g. c:\nxtcam_v1p1\nxtcam\i386

8. When the New Hardware Wizard pops up again, follow the same steps again.

Macintosh

Mac

1. Download the file FTDIUSBSerialDriver_v2_2_7.dmg from the USB Drivers area.

2. Double-click the file to begin the automated installation.

NXTCam Viewer

Now that you have your USB drivers installed for the camera, we can run a piece of software called NXTCam Viewer to see what the camera sees. This is an open source program hosted on Sourceforge, which means that if you want to add features to the software you can join the project and modify the code yourself. Windows users can download it from here:

http://nxtcamview.sourceforge.net

Macintosh users can download their own version from Mindsensors:

Mac

http://www.mindsensors.com

NOTE: Linux users are advised on the Mindsensors website to send them an email.

Install and run the NXTCam Viewer application. The first time you run the program, it will tell you the com port is invalid (see Figure 24-3). Use the drop down list (see Figure 24-4). If you don't see NXTCam there, make sure it is plugged in and your drivers are installed, then click refresh. Once you have it selected you can click test to see if it works.

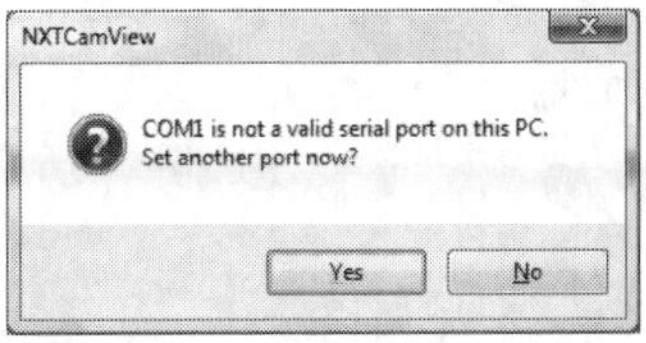

Figure 24-3: Getting an invalid com port error message

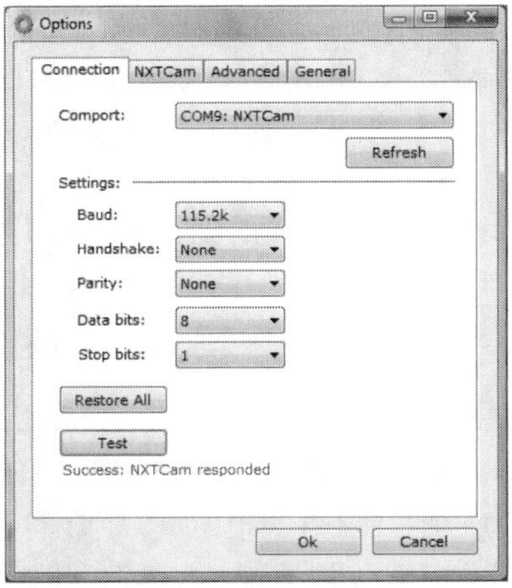

Figure 24-4: Selecting the NXTCam

Color Detection

The NXTCam uses color to track objects. It is up to you to tell the camera which colors to track. In this section we will set the color to detect some objects in the NXT kit. Click the Connect button and then click on Capture. Soon an image will appear on the screen (see Figure 24-5).

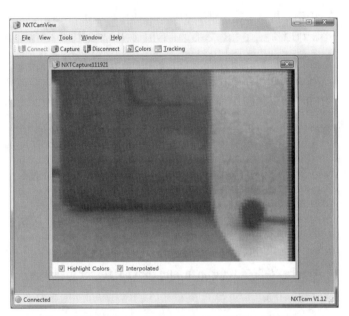

Figure 24-5: Capturing an image with the NXTCam

The NXTCam has limited ability to automatically adjust the colors for different camera exposure levels. Camera exposure refers to the lighting conditions in the room. Often a robotic project using a digital camera will work fine until the light level changes, then it stops working because colors look different under different light levels.

You should set the colors in the exact lighting conditions that your robot will operate in. If you set the colors in a part of the room with different lighting conditions, the colors will be set wrong when you move your robot to another part of the room, and it will not track the objects properly.

Make sure your area is well lit. If you have poor lighting, all objects will appear dark, no matter what color it is. This makes it hard for the camera to distinguish different colors.

One thing you will notice is that the image is very grainy because of the low 176 x 144 resolution. It isn't pretty, but it is good enough for basic object tracking. In fact, the actual NXTCam algorithm uses a reduced resolution of only 88x144 pixels in order to not overextend the camera controller. It will take a screen capture at 176 x 144, but it uses 88 x 144 for tracking.

To recognize a color, I used the red ball from the NXT 1.0 Kit, but you can use one of the Zamor spheres from the 2.0 kit as well. Once you have a screen capture, select the color icon from the toolbar (see Figure 24-6). This allows you to select a color in the image by moving the cursor around on the image and clicking the button to select a color. You'll see flashing yellow to represent colors that are detected.

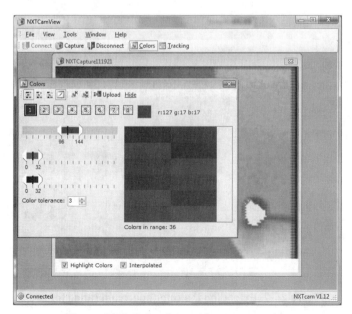

Figure 24-6: Selecting a color to recognize

Hopefully none of the colors in your background are recognized as the same color. If they are, you will see patches of the background flashing in yellow. This means parts of the background are the same color as your object, which is a potential problem.

You can adjust the color tolerance to recognize a wider or narrower range. You might want to try a few different angles and locations in your room to make sure the ball will be recognized no matter where it sits. It is very important to have evenly distributed lighting. Overhead lighting in the center of the room is the best. A few flashing dots in the background are not detrimental because we can weed those out later in the code by filtering out small rectangle sizes. You can also adjust individual color ranges (red, green or blue) by pulling either end of the color range. I stretched out the red range very wide because using a red ball (see Figure 24-7).

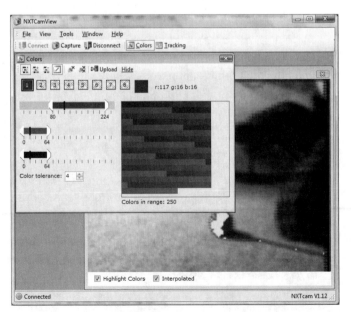

Figure 24-7: Extending the color spectrum range for red

You can test tracking within this application by clicking Tracking in the menu bar (see Figure 24-8). Once you click Start, you will see a live demonstration of tracking with the color you selected. My arm triggered the red color recognition, so I mounted the ball on a stick. Once it looks like the ball is recognized without confusing the background, click upload.

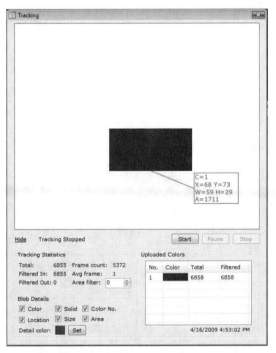

Figure 24-8: Tracking objects

You can also select multiple colors to detect. When selecting more colors (up to eight), you must click upload again with it selected to add that color to the NXTCam memory. The tracking window verifies there is more than one color.

Now that we have some colors set, try entering the following test code to output the position of objects to your NXT screen.

```java
import lejos.nxt.*;
import lejos.nxt.addon.*;
import java.awt.Rectangle;

public class NXTCamTest {

  final static int INTERVAL = 500; //
milliseconds
  public static void main(String[] args)
throws Exception {

    NXTCam camera = new NXTCam(SensorPort.S1);

    System.out.println(camera.getProductID());
    System.out.println(camera.
getSensorType());
    System.out.println(camera.getVersion());

    int numObjects;
    int counter = 0;

    camera.sendCommand('A'); // sort objects
by size
    camera.sendCommand('E'); // start tracking

    int OBJ_MIN = 6; // Minimum size

    while (!Button.ESCAPE.isPressed()) {
      System.out.println("Objects: " +
(numObjects = camera.getNumberOfObjects()) +
" counter = " + counter++);

      if (numObjects >= 1 && numObjects <=8) {
        Rectangle r = camera.getRectangle(0);
        if (r.height > OBJ_MIN || r.width >
OBJ_MIN) {
          System.out.print("X: " + r.x);
          System.out.println(" Y: " + r.y);
          System.out.print("Ht: " + r.height);
          System.out.println(" Wd: " +
r.width);
        }
      }
      Thread.sleep(INTERVAL);
    }
  }
}
```

Before running the code, plug the NXTCam into port 1. The camera will not work with the NXT brick while the USB cable is plugged into the computer. If you try to ask it how many objects there are, it will reply with -1. Make sure to disconnect the cable before running the NXT program.

Troubleshooting

1. Make sure you take the black plastic lens cap off the video camera before trying your robot. It's a small cap and easy to forget, but it will have you scratching your head wondering why your robot is failing to track objects properly.

2. Did you use NXTCamView to upload your color selection to the NXTCam?

3. Did the lighting conditions change due to an open window providing light? As the shun moves across the sky the color chosen at noon might look different through the camera at supper time.

4. Are there other colors in the room that are confusing the robot?

Predator Turret

In this section, we will build an automated turret that can hunt down paper targets. Unlike most projects in this book, this one uses parts exclusively from the NXT 2.0 kit, due to the Zamor Cannon included in this kit. You can order these parts from LEGO (see Table 24-1).

Just for fun!

The NXTCam can do line identification. Try using the Carpet Rover robot from earlier in the book. Mount the camera on the robot. Make a maze of straight lines and see if you can make your robot follow a maze.

Quantity	Part No.	Part Name
12	54821	Zamor sphere
2	53550	Zamor sphere magazine
1	54271	Zamor sphere launcher

Table 23-1: Acquiring the unique LEGO parts

In order to use this project, you will need to do these things in order to get it ready:

1. Build a turret robot
2. Set up the environment
3. Set up targets
4. Set color detection using NXTCam Viewer
5. Program the robot

Let's start with building the robot.

1

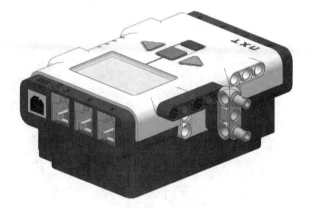

1x   1x

                  2x   1x

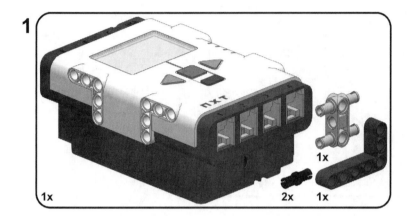

2

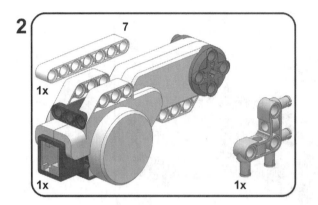

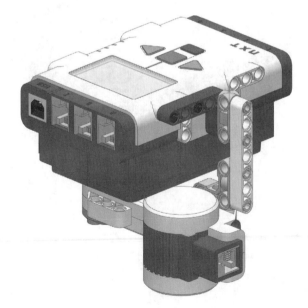

3

1x   7
1x   2x   1x

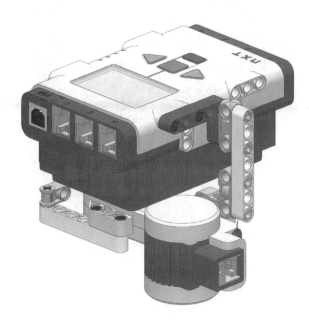

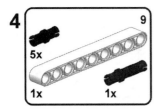

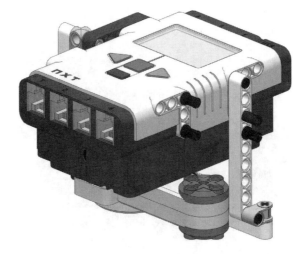

5

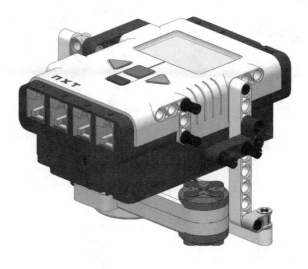

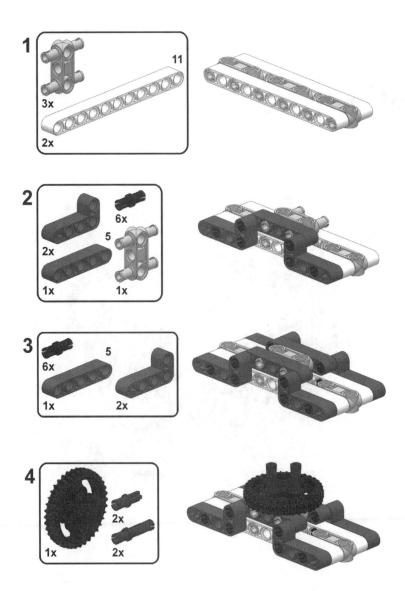

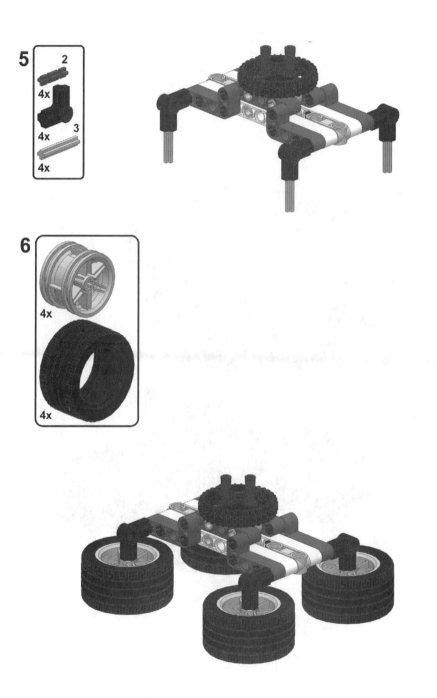

6

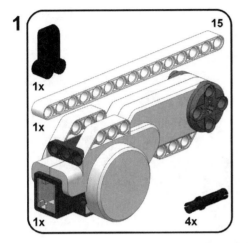

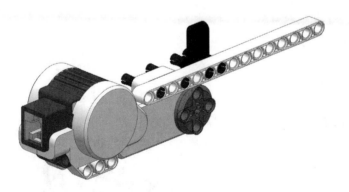

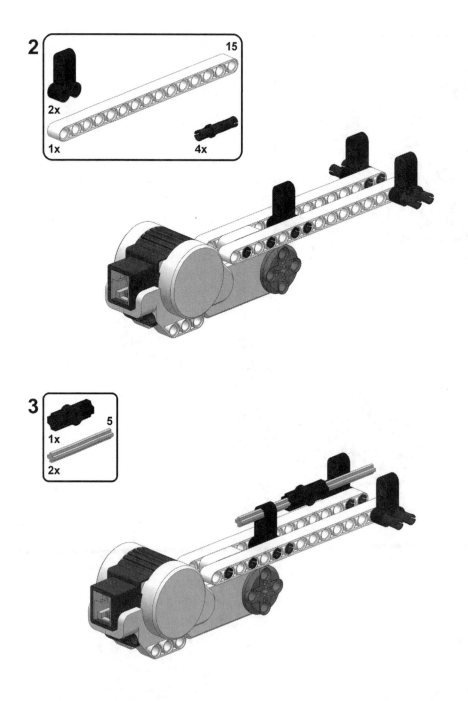

4

2x

1x

5

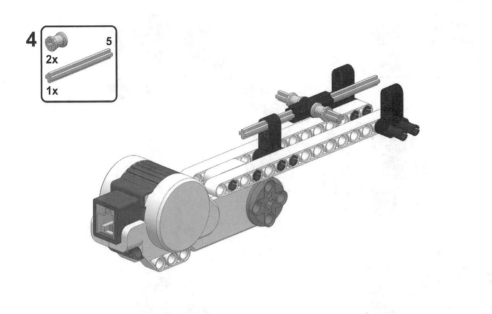

5

2x   2x

9

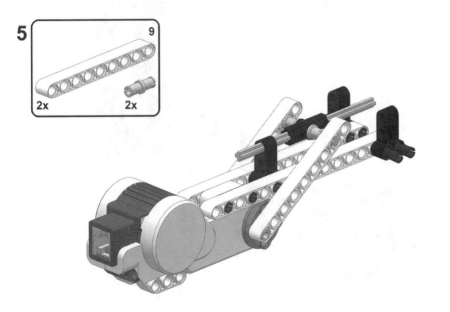

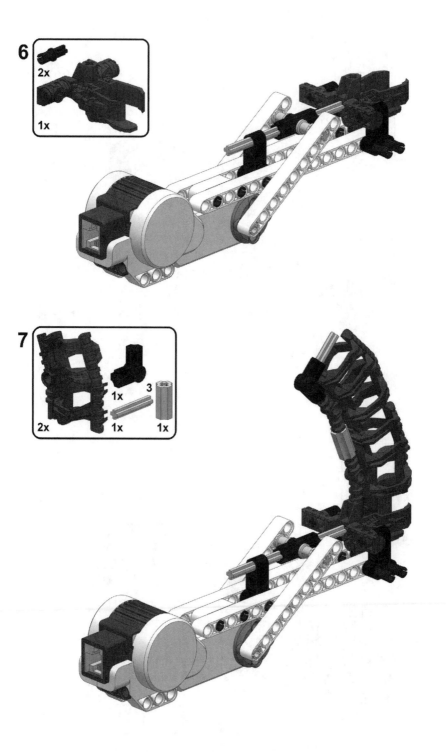

8

9

2x

2x   8x

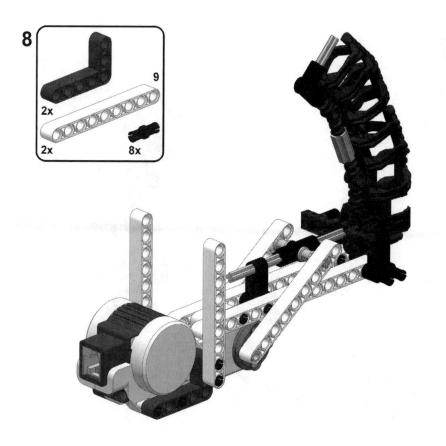

9

2x
2x
4x

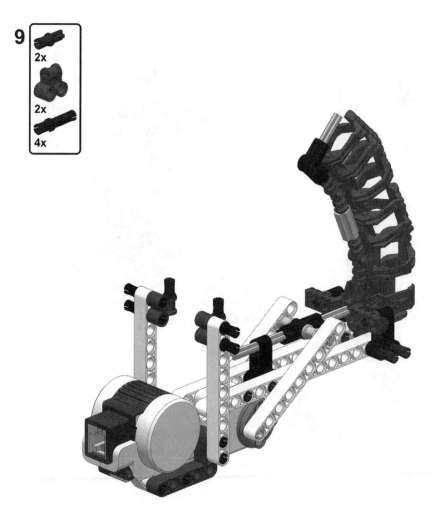

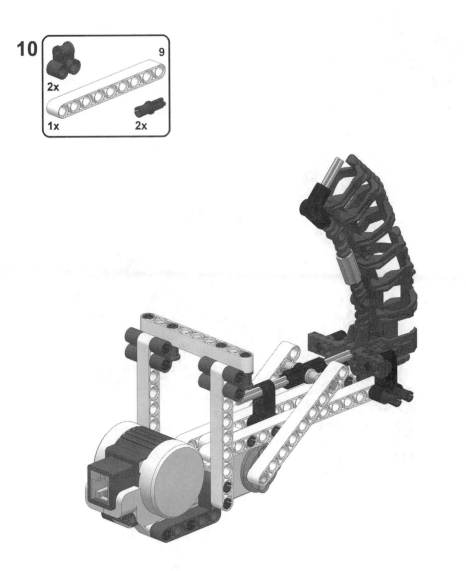

11

4x

2x

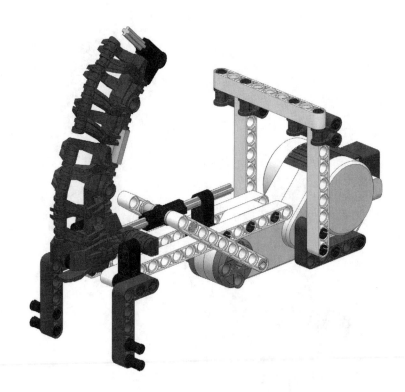

12

4x

5

2x

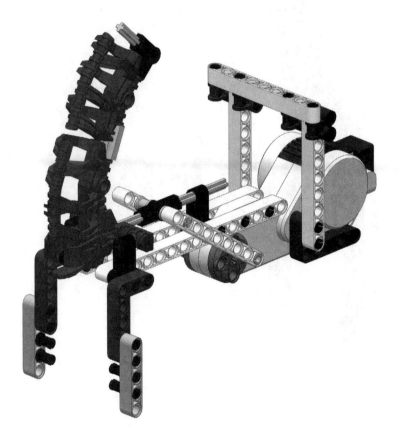

13

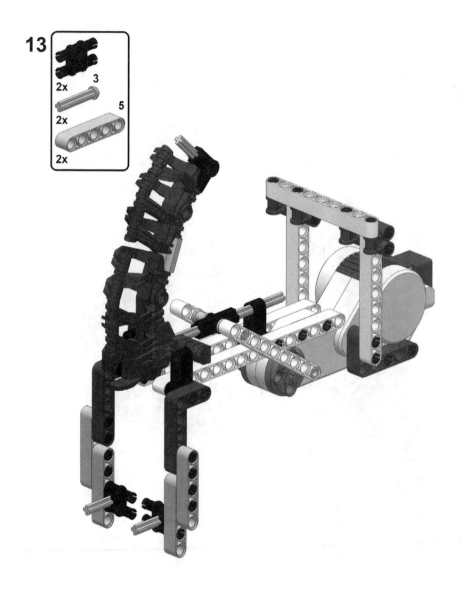

7

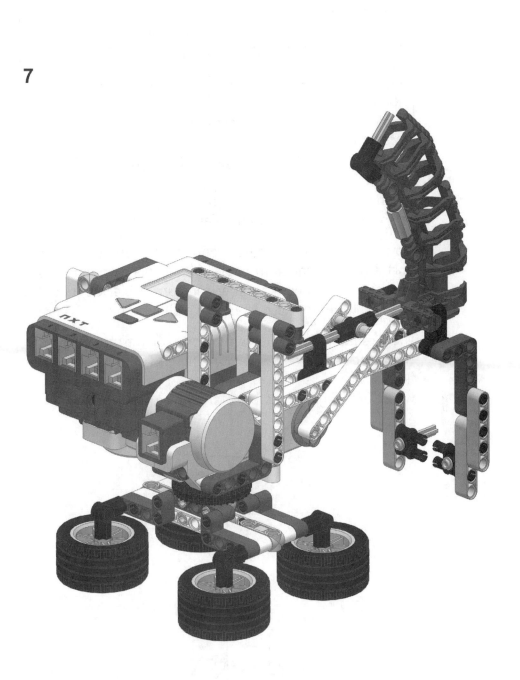

Turret Range	>6 meters (including roll)
Camera field of view	43 degrees
Minimum distance view	15 cm
Maximum distance view	150 cm

Table 24-2: Predator Turret Stats

When you are done building the model, mount the camera on the axles underneath the Zamor Cannon.

Now to plug in the cables. Connect the camera to sensor port 1 using a medium cable. Connect the cannon motor to motor port A with another medium cable. Connect the base motor to motor port B with a short cable.

The Predator Turret is reminiscent of the automatic turret in the movie *Aliens* (1986). The NXTCam has a limited field of view as shown in Table 24-2, but the turret covers a whole 360 degree field by rotating (see Figure 24-9).

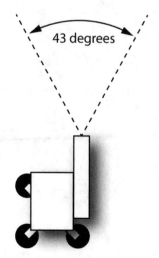

Figure 24-9: Field of view of the turret camera

Programming Predator Turret

The turret code needs to do these things in order:

1. Scan: Rotate until a target is acquired in the field of view.

2. Aim: Rotate until the target is in the center of view.

3. Fire: Launch the projectile at the target

```java
import lejos.nxt.*;
import lejos.nxt.addon.*;
import java.awt.Rectangle;

public class Predator {

  final static int INTERVAL = 200; //
milliseconds
  final static int FIELD_OF_VIEW = 52; //
degrees camera covers
  final static int PIXEL_WIDTH = 176; //
camera width
  final static int CAMERA_CENTER = PIXEL_
WIDTH/2; // camera width

  final static int SCAN_ZONES = 10; // zones
to scan

  public static void aim(Rectangle r) {
    int centreX = r.x + (r.width/2);
    int x_diff = centreX - CAMERA_CENTER;
    float x_ratio = (float)x_diff / PIXEL_
WIDTH;
    float rotateDeg = x_ratio * FIELD_OF_VIEW;
    Motor.B.rotate((int)-rotateDeg);
  }

  public static void rotateToZone(int zone) {
    Motor.B.rotateTo(-zone * (FIELD_OF_VIEW -
10));
  }

  public static void main(String [] args)
throws Exception {
    Motor.B.setSpeed(50);
    NXTCam cam = new NXTCam(SensorPort.S1);
    int numObjects;
```

```
cam.setTrackingMode(NXTCam.OBJECT_TRACKING);
cam.sortBy(NXTCam.SIZE);
cam.enableTracking(true);

int zone = 0;

while(zone < SCAN_ZONES) {
  rotateToZone(zone);
  numObjects = cam.getNumberOfObjects();

  if (numObjects >= 1 && numObjects <= 8)
{

    Sound.beep();
    Rectangle r = cam.getRectangle(0);
    Predator.aim(r);
    Motor.A.rotate(360);
  }

  ++zone;
  Thread.sleep(INTERVAL);
  }
 }
}
```

Before we can use this program we need to build some targets and set the NXTCam to recognize the colors of those targets.

Building Targets

In this section you get to pull out some scissors and colored cardboard. I recommend orange, but basically any color will do as long as it is not similar to the floor and walls of your room. Preferably you should use a color on the opposite side of the color wheel from your room (see Figure 24-10).

Do you have blue walls? Use yellow targets. Do you have green walls? Use purple. Doing so will provide the NXTCam with a high contrast (a larger separation from other colors) which will help it to recognize the targets.

The target material should be heavier than regular paper because they will be hit with high-velocity zamor spheres. You can find colored construction paper at almost any office supply store.

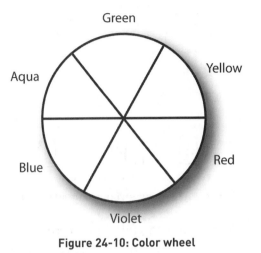

Figure 24-10: Color wheel

The targets are formed into a special shape for maximum recognition by the NXTCam. The face of the target that is visible to the camera cannot be a flat surface because the colors change radically depending on the location of the targets relative to the light (see Figure 24-11).

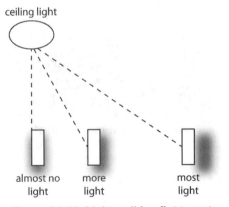

Figure 24-11: Light striking flat targets

To combat this, the top of the target has a curved surface (see Figure 24-12). This allows light from the ceiling to strike all the targets evenly, and it spreads out the range of shades so that each target likely contains at least a thin band of the color the camera is trying to detect.

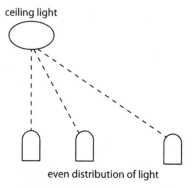

ceiling light

even distribution of light

Figure 24-12: Light striking curved targets

1. Cut the cardboard into 20 by 10 cm rectangles.

2. Fold over and tape/glue the bottoms (see Figure 24-13).

3. If you want to get fancy, use white paper on the bottoms so that when it is knocked over the camera can't see the color anymore.

4. It is easy for the camera to detect close targets because they are larger in the frame, but you will need larger targets if they are more distant. Make some larger targets for farther ranges.

Figure 24-13: The curved targets

Setting Color Tracking for Targets

Place the Predator Turret in middle of your test area. Again, make sure the wall colors are not the same as the target colors. Clear away similarly colored objects in the background.

Set up the targets around the robot as if you were really going to run the program. Now we will use the camera and your PC to look at each target to see what it looks like to the computer.

It is important to use the actual setting and lighting conditions in which the turret will operate. It might seem like the robot will work in another room, when in fact it probably won't.

Now use the NXTCam Viewer to set the color, as shown earlier in this chapter. When choosing a color, make sure to click on the top part of the target where it is curved. As previously mentioned, the flat face of the target is problematic because it is susceptible to lighting conditions making the same color appear to be a different shade.

Next we will try tracking the target to make sure the camera recognizes it. Notice sometimes it picks out 2 or even 3 bounding boxes for one single object? In order to prevent the robot from firing multiple times at the same object, we can just ignore the smaller rectangles.

Running the program

Once the colors are uploaded to the NXT-Cam, unplug the USB cable and run the program. It should rotate around the targets, take aim and fire. There is no guarantee it will hit each target, however.

Just for fun!

Try mounting the Predator Turret on a mobile robot. The turret will no longer require the rotating base because the entire robot can rotate to aim the turret. Now your robot can become a relentless paper-target hunting machine!

CHAPTER 25

RFID

TOPICS IN THIS CHAPTER

- RFID Sensor and Transponder
- leJOS Java API
- Pet food dish

Radio-frequency identification, commonly known as RFID, is a method to identify an item using radio frequencies. Chances are that you have shopped at a store that used RFID tags on clothing to prevent theft and identify items at the checkout. In the livestock industry, farmers often tag the ears of animals with an RFID tag. Not only are they used to identify individual animals but they can also monitor how much each animal eats. Every time the animal steps up to the feed bin, the RFID sensor records the feeding time and (using a digital scale) how much food the animal consumes. There are numerous uses for RFID with LEGO NXT and this chapter explores some of them.

LEGO RFID Sensor

There are two parts to RFID, the sensor and the transponder. When a transponder is within range of the radio receiver (RFID sensor), it receives a radio signal from the transponder and reads in a unique identification number.

There are two kinds of transponders, active and passive. Active transponders contain a battery to power the radio signal. Passive transponders rely on an external electromagnetic field to provide electricity for the transponder chip, usually provided by the RFID sensor. Because passive transponders do not require a battery, they are much cheaper to manufacture and therefore far more popular in the marketplace.

> **NOTE:** You can find lots of RFID supplies on eBay, such as receivers and transponders. If you want to use them with your LEGO RFID sensor, make sure to buy 125 kHz transponders.

LEGO Education sells an RFID kit for the NXT. The small box contains an I²C sensor (see Figure 25-1) and numerous transponders. This kit uses passive transponders, so the sensor creates a weak electromagnetic field which activates the transponders when they are in the vicinity. Due to battery considerations, the electromagnetic field is weak and therefore only good for distances up to three centimeters.

Figure 25-1: Viewing the RFID Sensor

The RFID kit comes with a sensor and a dozen transponders with unique ids. There are five transponder key-tags, five clear transponder 'coins', and two white 'credit card' transponders (see Figure 25-2). The clear discs are interesting because you can see the coil which converts an electromagnetic field into electricity and two thin wires leading to the chip. Likewise you can see through the clear-orange sensor cover to see the copper wire wound coil and black cylindrical magnet which generates the electromagnetic field (see Figure 25-1).

RFID is potentially useful for overcoming one of the most important robotics problems, object recognition. Since it is very complicated for robots to identify objects visually, an RFID tag attached to objects in the environment can shortcut the problem and allow positive identification.

Website:

For more information about the RFID Sensor, programming examples, and to download the programming block, visit the manufacturer website.

www.codatex.com

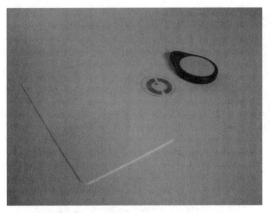

Figure 25-2: Viewing the transponders

Programming RFID

The lejos.addon package contains a class called RFIDSensor. This class allows you to read data from your sensor. There is really only one practical method in the RFIDSensor, which is used to read a value from an RFID transponder.

- long readTransponderAsLong(boolean continuous)

If *true* is used as a parameter then the sensor will continue reading and will record any values found between this read and the next time this method is called. If you just want a one-shot read, use false.

Just for fun!

Wikipedia has a comprehensive list of various ways to use RFID, such as keyless entry. The list might give you some ideas for your own projects.

http://en.wikipedia.org/wiki/RFID

Pet Food Protector

Now that we are familiar with RFID and how to take readings, let's try out a fun project. This project was inspired by observing that some dogs try to steal food from cats or smaller/slower dogs. This project will ensure that each pet only eats from its own bowl.

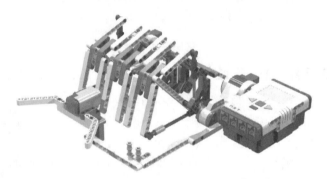

Figure 25-3: The RFID pet food protector

Building the Pet Food Protector

The LEGO NXT kit contains enough parts to form a cage that will protect a food dish from dog noses, but it will take quite a few parts.

> **NOTE:** If you are using an NXT 2.0 kit below, you will need to substitute the two 8-unit axles with two 9-unit axles.

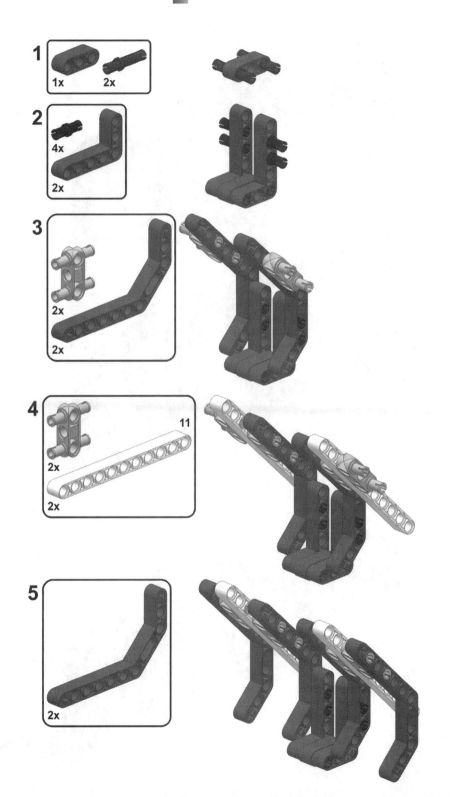

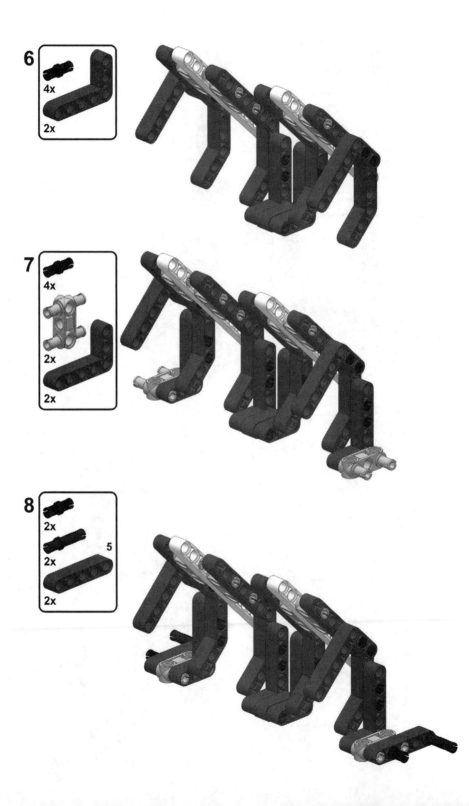

9

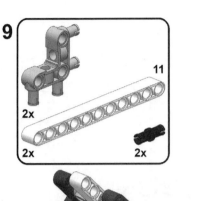

11

2x

2x   2x

10

5   11

1x

2x   4x

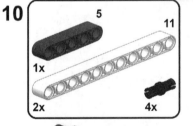

1

1x

4x

2x

4x

2

4x

7

2x

9

2x

3

8x

7

4x

9

4x

2x

4x

8x

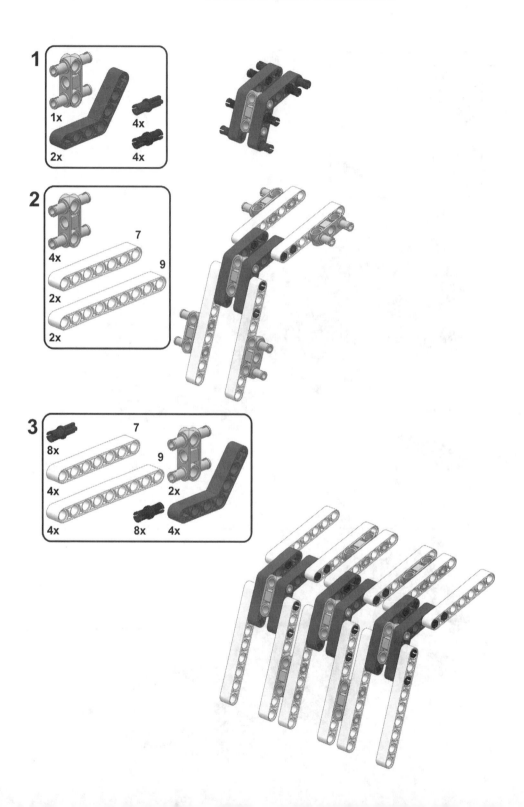

4

2

2x

2x

2x

5

8

2x

2x   2x

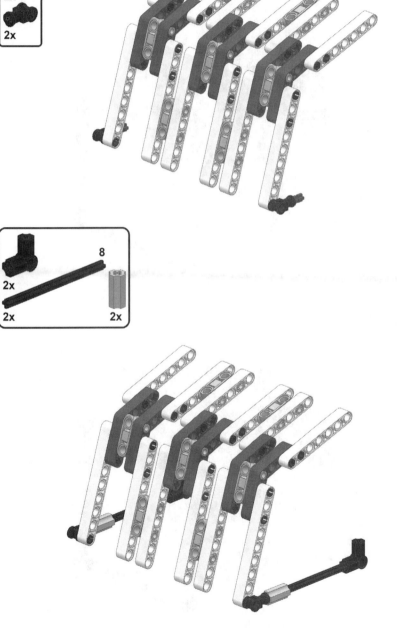

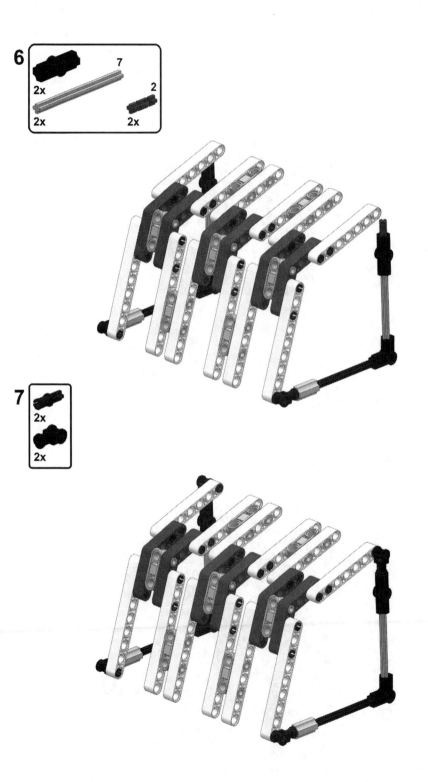

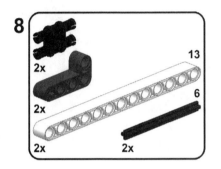

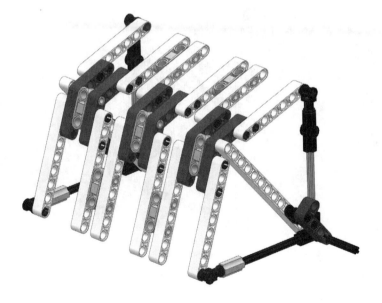

11

2x

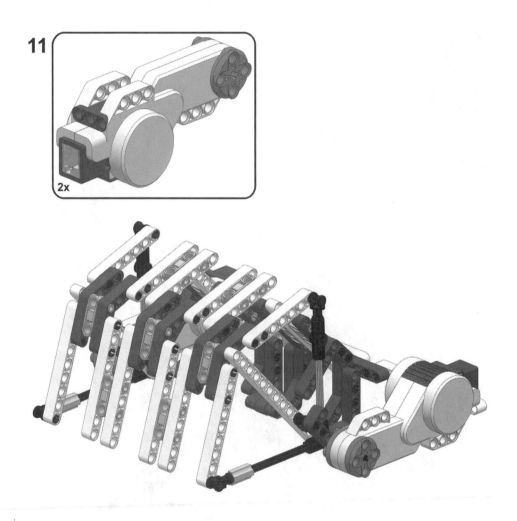

12

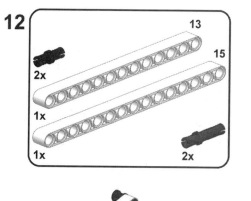

13
15
2x
1x
1x
2x

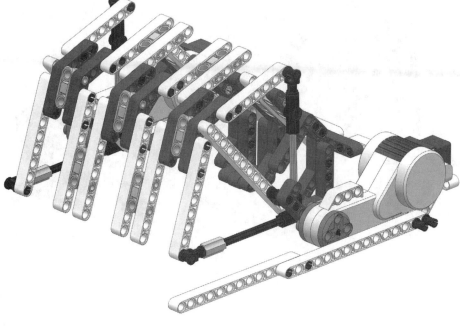

13

2x

1x

14

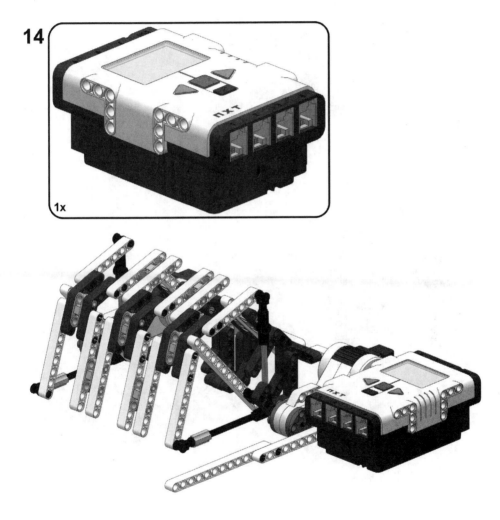

1x

15

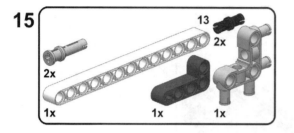

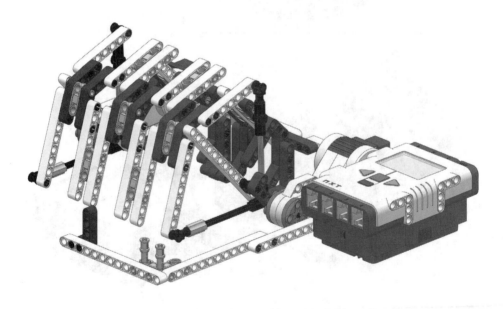

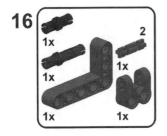

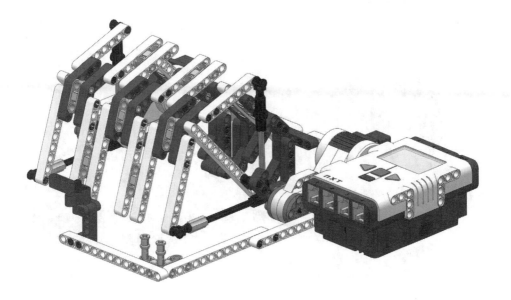

17

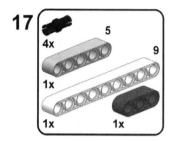

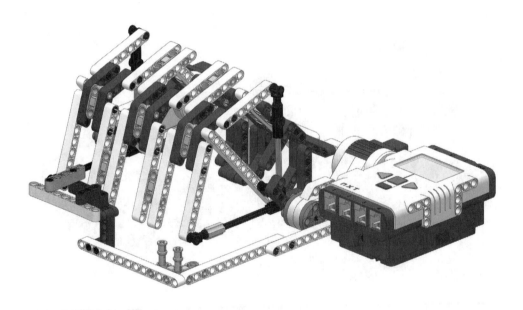

18

1x
2x
2x   4x

Use a medium cable to connect the RFID sensor to port 1. Wind the cable through the two bush-pins at the corner of the frame to keep the cable away from the cage (see Figure 25-4). Using a short cable, connect port B with the closest motor. Use a medium cable to connect the remaining motor to port C.

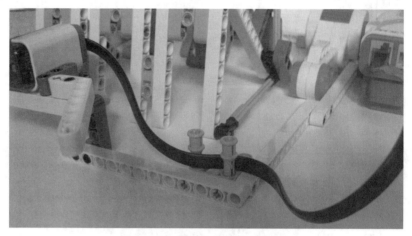

Figure 25-4: Wrapping the cable

Programming the Food Protector

The control logic for the pet food protector is simple. It merely opens the dish if it detects the correct transponder in the vicinity. If the tag is not detected within a certain period of time it closes the dish protector.

```java
import lejos.nxt.*;
import lejos.nxt.addon.RFIDSensor;

public class RFIDPetDish {

  public static int CLOSING = 0;
  public static int OPENING = 1;
  public static int T_STAY_OPEN = 30000; //
Delay before closing (millis)

  public static void main(String[] args) {
    int state = CLOSING;
    int oldState = CLOSING;
```

```
    Motor.B.setSpeed(50);
    Motor.C.setSpeed(50);
    RFIDSensor rfid = new
RFIDSensor(SensorPort.S1);
    System.out.println(rfid.getProductID());
    System.out.println(rfid.getSensorType());
    System.out.println(rfid.getVersion());
    System.out.println("Put tag in front\nof
sensor and \npress enter");
    Button.ENTER.waitForPressAndRelease();
    long tagNumber = rfid.
readTransponderAsLong(false);
    Sound.beepSequenceUp();
    long time = 0;
    while(!Button.ESCAPE.isPressed()) {
      long transID = rfid.
readTransponderAsLong(true);
      if(transID == tagNumber) {
        System.out.println(transID);
        time = System.currentTimeMillis();
      }
      long curTime = System.currentTimeMillis();
      if(curTime - time > T_STAY_OPEN)
        state = CLOSING;
      else
        state = OPENING;

      if(state != oldState) {
        oldState = state;
        if(state == CLOSING) {
          Sound.beepSequence(); // Warning noise
          try {Thread.sleep(1500);} catch
(InterruptedException e) {}
          Motor.B.rotateTo(0, true);
          Motor.C.rotateTo(0, true);
        } else if(state == OPENING){
          Sound.beepSequenceUp();
          Motor.B.rotateTo(-90, true);
          Motor.C.rotateTo(-90, true);
        }
      }

      try {Thread.sleep(100);} catch
(InterruptedException e) {}
    }
  }
}
```

Using the Pet Food Protector

Before running the code, make sure the dish is inside the cage and the cage is closed shut. When you first run the program it will ask you to place the transponder next to the sensor and press enter. Go ahead and do this. Once it beeps, you can remove the transponder and put it on your pet. The dish protector is ready to go!

This is a fun project because it involves unpredictable animals. It would have been an easier project if the RFID sensor could detect transponders at a longer range. Due to the three centimeter limit, it was essential to situate the sensor at the proper height and distance, and even to try to funnel the transponder over the sensor.

I fixed the transponder key tag onto my cat's collar with a thin piece of wire (see Figure 25-5). The transponder hangs down which puts it in a good position to trigger the dish. I had to lure my cat to the correct position using bits of food placed on the RFID sensor. The good news is that food is a great motivator for animals to learn new tricks. She tried pulling cat food out with her paw, but eventually learned it is easier just to walk up to the front of the dish.

Just for fun!

For more RFID ideas, the United States patent office is a good place to start. From the main page, select Search › Quick Search and use the term RFID. You can literally browse through thousands of inventions involving RFID sensors. Many of them are interesting projects you could try out with your LEGO NXT and the RFID kit.

www.uspto.gov

Figure 25-5: Attaching the transponder to a collar

CHAPTER 26

Monte Carlo Localization

TOPICS IN THIS CHAPTER

Landmark navigation is when you look at objects in the environment to pinpoint your location. As chapter 6 mentioned, sailors often used islands and other landmarks to pinpoint their location. GPS is another form of landmark navigation, using the known position of satellites. In this chapter, we will use a different form of landmark navigation known as Monte Carlo Localization to pinpoint location.

Global Localization

For many navigation tasks, the starting point of your robot is known and the problem is keeping track of it while it moves around. This is known as a tracking problem, and it is the easiest of the localization problems. The previous chapters introduced this concept.

But what if you do not know the starting point of the robot? Is there a way that the robot can find out where it is by taking sensor readings from its environment? There is, and it is known as *global localization*.

The robot could use a GPS sensor to find out where it is, but there are several problems with this approach. First, you might not have a GPS sensor, and you might want to solve the global localization problem with only the sensors that come with your LEGO NXT kit. Second, a GPS sensor typically only gives you the robot's position, but to locate a robot and keep track of it while it moves, you also need to know the heading of the robot. A standard GPS sensor will not tell you that. A third problem is that a GPS sensor may not be accurate enough. You will typically want to know the position of your robot to the accuracy of a few centimeters and a standard GPS sensor is only accurate to several meters. Finally, you might be indoors, out of view of satellites, and might not be able to get a GPS signal.

There are two variants of the global localization problem depending on whether the robot has a map of its environment.

We will tackle the problem where the environment has already been mapped (see Chapter 13). The harder problem of Simultaneous Localization and Mapping (SLAM) is the subject of current research in robotics.

The easier problem of global localization using a map is still a difficult problem. We will use a technique called Monte Carlo Localization (MCL) to solve this problem.

Monte Carlo Localization

Imagine you are kidnapped and wake up in a strange environment, such as a large mansion full of rooms and corridors (see Figure 26-1). You obviously want to get out, but you do not know which way to go. You find a map of the mansion that shows the walls of all the rooms and corridors (see Figure 26-2). You feel hope surge through your body as you spot the exit, but of course it does not show where you are. How would you use this map to find your way out? After all, you don't know your location so you have no starting point to plan your exit.

Figure 26-1: Finding your way out of a mansion

You could look at the room you are in and compare it with rooms on the map (see Figure 26-2). If you can match the geometry of the room to one of the rooms on the map, you would know exactly where you are. However, in a large complex mansion there may be many similar rooms and corridors, so there might be several different possibilities for where you are. You would need to keep track of the various possible locations and, as you moved about and got more information, you would discard some of the possibilities. With any luck you would eventually be left with just one possibility. You would then know your location and could use the map to quickly make your way to the exit.

Figure 26-2: Examining the map for an escape route

This is roughly how the Monte Carlo Localization algorithm works, but the robot does not have sensors as sophisticated as your eyes. It has a range sensor and can take multiple range sensor measurements to the nearest walls. The robot must compare these measurements with the measurements it would get if it were in the different possible locations in the map and use this information to calculate the most probable of all the possible locations.

The algorithm is called Monte Carlo because it relies on probability like the spins of a roulette wheel in the casinos of Monte Carlo.

Can a robot really find its location with just a map and a range sensor? Yes it can, along with the localization API in leJOS NXJ. Using the API, your program can be relatively simple. It might occupy only 50 lines. However, behind the scenes there are over 1000 lines of code to make the algorithm work.

Map

The problem we are going to solve is to find the position of a mobile robot in a two-dimensional environment such as a walled room. We will not try anything as complex as a large mansion with many rooms and corridors, but the principles are the same.

The robot will take sensor measurements to try to determine its pose. The ultrasonic sensor can take range measurements from the robot to the nearest object, which could be a wall or other obstacle. In this example, we will limit ourselves to bare rooms where the only obstacles are walls, but a map can represent other obstacles as long as they are represented by straight line segments.

We need to be able to compare the range readings obtained by the robot with the theoretical readings of many different possible poses. The Monte Carlo algorithm will use the RangeMap.range() method to determine the distance from a pose to the nearest wall or other obstacle in the map.

The environment we will use to test the MCL algorithm is a simple walled area of a room (see Figure 26-3). This map has all perpendicular walls, but you are not limited by this for MCL. You should use the SVG map you made in chapter 13 instead of the example map in this chapter.

All measurements in the map are in centimeters. Note that y values increase from top to bottom.

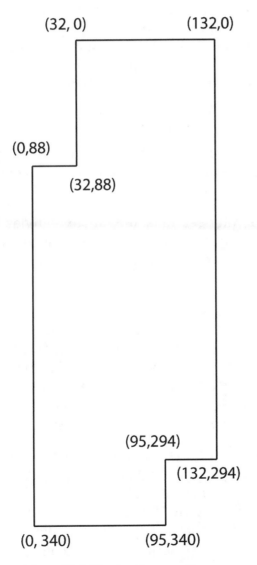

Figure 23-3: Viewing the example map

NOTE: In order for the MCL algorithm to work properly, the map must not be symmetrical or mirrored. In other words, a square or rectangle will not work because the MCL algorithm will be unable to pinpoint a unique location. The mapped area must have some defining features for the algorithm to identify position.

Locating the Robot

Now that we have a map, a robot and a test area, we can place the robot in a random position in the test area and ask it to find out where it is. Without any special software, the robot is unlikely to answer. We will need to tap into the powerful MCL classes in the localization package.

If we take a single ultrasonic reading there are many poses that the robot could have that are consistent with this reading. For example, if the robot receives a reading of 50 cm, it could be 50 cm from any wall with any orientation. This does not tell us very much.

Instead, we will need to take multiple readings in order to determine unique features of the surroundings. The ultrasonic sensor is not very accurate above about 150 cm, so if our mapped area is larger than this, we cannot take measurements to all the walls.

It would be nice to have multiple sensors to take simultaneous readings in multiple directions, but unfortunately we only have one ultrasonic sensor. We could mount the ultrasonic sensor on a turntable and take multiple readings, but a simpler solution is to just rotate the robot to take multiple measurements and then rotate it back to its original pose.

We will take three measurements: one forward and one each at 45 degrees to the left and right. This will identify the pose of the robot with respect to the walls in front of it much more accurately than a single reading. In Figure 26-4a, a pose (which shows three range readings) fits the map data near the corner perfectly. The second pose in Figure 26-4b does not match the geometry around it.

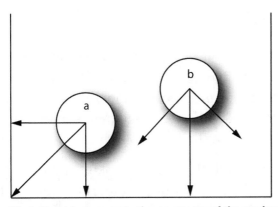

Figure 26-4: Matching poses to the geometry of the environment

Even with this set of readings, there are still many poses that the robot could have that are consistent with these readings. As we did when escaping from the mansion, we will have to move about and examine lots of geometry to establish a location with any confidence.

But we also have another problem. Ultrasonic sensor readings are not very precise and often include random noise. The cone shape of the sonar pulse, as described in Chapter 4, is not as accurate as a laser. Therefore if it is at an angle to a wall it might return the value where the cone first comes into contact with the wall (see Figure 26-5). Or more likely, the farthest part of the cone will be returned while the closer one will produce a glancing reflection that does not return to the ultrasonic sensor.

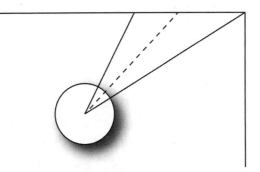

Figure 26-5: Inaccurate readings due to the sensor cone

Particle Set

Despite having only a set of approximate readings, the MCL algorithm can accurately predict location. To do this, it needs a sensor probability model that tells it how probable it is that a set of three readings (see Figure 26-4) are produced by any particular pose.

Things now get a little more complex, and perhaps a little intimidating. At the start of the algorithm, before the robot has made any moves, the MCL algorithm generates a whole bunch of random poses all over the map (see Figure 26-6).

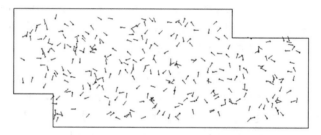

Figure 26-6: A random particle set

These are a sampling of all the possible locations of the robot. The range sensor then takes a set of three range readings. Using those readings, the MCL algorithm knocks out all the random poses that are not possible with the observed readings and the given probability model (see Figure 26-7). The remaining poses are all possibilities of the current location.

If MCL relied on exact measurements, this algorithm would not work. The ultrasonic sensor has error, as mentioned above. The map data invariably has error. Using exact measurements, the algorithm would conclude that none of them matched up exactly as it should have. Instead, we use statistical probability. So even though the map might not be accurate, MCL still finds the most probable location.

Figure 26-7: Disqualifying most of the random particles

The robot then makes an actual physical move by rotating and travelling in some direction. Random noise is added to each move proportional to the distance moved or the angle turned. This means that particles that start out in the same position will spread out and form little clusters after each move is made (see Figure 26-8). So, what looks like a single particle after pressing Get Pose, can actually be multiple particles in exactly the same position.

Each of these particles has the random move applied to it, and will get a different random amount of noise. So what looked like a single particle will turn into a small cluster with each particle having a slightly different position and heading, after the random move. This actually helps to increase accuracy, given that movement from a robot is never exact. By having distinct particles with different positions and headings at the end of each move, one of them is sure to be closer to the actual pose.

Figure 26-8: Generating a new random particle set

It then takes another range scan which is uses to further knock out some of the random particle set. After a few moves, hopefully it has narrowed down the particle set to a tight cluster, which indicates it knows with certainty its location on the map (see Figure 26-9). The theory is straightforward but the results are extraordinary.

Figure 26-9: A tight cluster of particles indicating success

Monte Carlo Localization in Action

That's enough theory for now. Let's try coding a visual example of MCL to see how it works in practice. In order to see the MCL algorithm performing real-time localization, it is necessary to display the data to a computer screen (the LCD screen on the NXT is very small and also moves with the robot, making it difficult to view).

Enter the following program into a leJOS PC Project in Eclipse (see chapter 18 for more information on PC projects).

> **NOTE:** As of this writing, the mapping classes are new to leJOS and might undergo more revisions. If you find this code is out of date, try using the latest code MCLTest located in the pcsamples and samples directory.

```
import java.awt.*;
import java.awt.event.*;
import javax.swing.*;
import lejos.robotics.mapping.*;
import lejos.robotics.mapping.
NavigationModel.NavEvent;
```

```java
import lejos.robotics.localization.*;

public class MCLDemo extends NavigationPanel
{
  private static final long serialVersionUID
= 1L;

  // GUI Window size
  private static final int FRAME_WIDTH =
1000;
  private static final int FRAME_HEIGHT =
800;
  private static final int NUM_PARTICLES =
200;

  private static final JButton randomButton =
new JButton("Random move");
  private static final JButton getPoseButton
= new JButton("Get Pose");
  private static final String mapFileName =
"MyRoom.svg";

  private static MCLPoseProvider mcl;

    /**
    * Create a MapTest object and display it
in a GUI frame.
    * Then connect to the NXT.
    */
    public static void main(String[] args)
throws Exception {
      (new MCLDemo()).run();
    }

    public MCLDemo() {
      buildGUI();
    }

    protected void buildGUI() {
      // All panels required
      super.buildGUI();

      // Add the Get Pose and Random Move
buttons
      commandPanel.add(getPoseButton);
      commandPanel.add(randomButton);

      // disable buttons until connected
      getPoseButton.setEnabled(false);
      randomButton.setEnabled(false);
```

```
    // When Get pose is pressed, invoke the
MCL Pose provider
    // to take readings and get the pose. Then
get the updated
    // particles, the details of the estimated
pose and the range
    // readings.
    getPoseButton.addActionListener(new
ActionListener() {
      public void actionPerformed(ActionEvent
event) {
        model.getPose();
        model.getRemoteParticles();
        model.getEstimatedPose();
        model.getRemoteReadings();
        getPoseButton.setEnabled(false);
      }
    });

    // When the Random Move button is pressed,
make a random move
    // and get the updated particles
    randomButton.addActionListener(new
ActionListener() {
      public void actionPerformed(ActionEvent
event) {
        model.randomMove();
        model.getRemoteParticles();
      }
    });
  }

  /**
   * Called when the mouse is clicked in the
map area
   */
  protected void popupMenu(MouseEvent me) {
    // Calculate the screen point and create
the pop-up the context menu
    Point pt = SwingUtilities.
convertPoint(me.getComponent(),
me.getPoint(), this);
    JPopupMenu menu = new JPopupMenu();

    // Get details of the particle closest
to the mouse click
    menu.add(new MenuAction(NavEvent.FIND_
CLOSEST, "Find Closest", me.getPoint(),
model, this));
```

```
    // Show the context menu
    menu.show(this, pt.x, pt.y);
  }

  /**
   * Called whenever an event is received from
the NXT
   */
  protected void eventReceived(NavEvent
navEvent) {
    // Enable the Get Pose button when the
estimated pose has been sent
    if (navEvent == NavEvent.ESTIMATED_POSE) {
getPoseButton.setEnabled(true);
    }
  }

  /**
   * Called when the connection is established
   */
  protected void whenConnected() {
    // Load the map and generate the particles
and sends both to the NXT
    model.loadMap(mapFileName);
    model.generateParticles();

    // Enable buttons
    getPoseButton.setEnabled(true);
    randomButton.setEnabled(true);
  }

  /**
   * Run the sample
   */
  public void run() throws Exception {
    // Create a stub version of the
MCLPoseProvider
    mcl = new MCLPoseProvider(null,NUM_
PARTICLES,0);

    // Associate the MCLPoseProvider with the
    model model.setMCL(mcl);

    // Open the MCLDemo navigation panel in a
JFrame window
      openInJFrame(this, FRAME_WIDTH, FRAME_
HEIGHT, "MCL Test", Color.white);
  }
}
```

This program will load in your map data from a file called My-Room.svg. Use the file you made in chapter 13 and place it into the same directory as the code. If you are using Eclipse, this is simple. Just copy the file from your directory by right-clicking the file and selecting copy. Then paste the file into the Eclipse project by right clicking the project and selecting paste.

You can run the program on the PC, but don't try connecting or loading the map until we have the corresponding code running on the NXT robot.

Note that the class NavigationPanel, which draws the map and particles and sends commands to the NXT, is included in the lejos.robotics.mapping package. You can use this class to make your own custom mapping application.

Programming the NXT

As previously mentioned, a program runs on the NXT to communicate with the PC. To exchange data, the NXT uses the NXTNavigationModel class.

```
import lejos.nxt.*;
import lejos.robotics.*;
import lejos.robotics.localization.
MCLPoseProvider;
import lejos.robotics.mapping.
NXTNavigationModel;
import lejos.robotics.navigation.*;

public class NXTMCL {
  private static final float[] ANGLES =
{-45f,0f,45f};
  private static final int BORDER = 0;
  private static final double ROTATE_SPEED = 100;
  private static final double TRAVEL_SPEED = 100;
  private static final float MAX_DISTANCE = 40f;
  private static final float PROJECTION = 20f;

  public static void main(String[] args)
throws Exception {

    DifferentialPilot robot = new
DifferentialPilot(5.6, 16.4,Motor.B,Motor.C);
    robot.setRotateSpeed(ROTATE_SPEED);
```

```
    robot.setTravelSpeed(TRAVEL_SPEED);
    robot.setAcceleration(500);
    RangeFinder sonic = new
UltrasonicSensor(SensorPort.S1);
    RangeScanner scanner = new
FixedRangeScanner(robot, sonic);

    scanner.setAngles(ANGLES);
    // Map and particles will be sent from
the PC
    MCLPoseProvider mcl = new
MCLPoseProvider(robot, scanner, null, 0,
BORDER);
    Navigator navigator = new Navigator(robot,
mcl);
    NXTNavigationModel model = new
NXTNavigationModel();
    model.setRandomMoveParameters(MAX_
DISTANCE, PROJECTION, BORDER);
    model.addPilot(robot);
    model.addNavigator(navigator);
    model.addPoseProvider(mcl);
    model.setAutoSendPose(false);
  }
}
```

This example will use the Carpet Rover from earlier in the book. The motors are plugged into ports B and C. You will need to attach an ultrasonic sensor and plug it into port 1. To improve the accuracy of sensor readings, it is important to mount the ultrasonic sensor directly above the centre of rotation of the robot (see Figure 26-10).

Figure 26-10: Centering the ultrasonic sensor

Exploring the environment

Before running the program, you need to change the map to correspond to your walled area. The area needs to be a few square meters in size with straight walls and no un-mapped obstacles. The robot should have clear views of the walls at the height of the ultrasonic sensor.

Place the robot in the mapped area and start the NXT program. The robot is best placed so that it is pointing at walls within range of the ultrasonic sensor. It will wait for a connection with the PC.

Now run the PC program from Eclipse. Enter the name of your NXT brick and click the connect button. The PC program sends a command to the NXT to show the initial particles (see Figure 26-11).

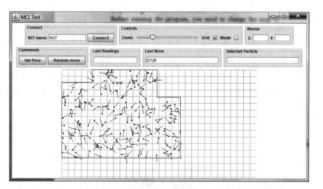

Figure 26-11: The Randomly generated particles

Now press Get Pose and the robot will immediately take three range readings from the environment and calculate the weights of each particle. This takes some time, so be patient. You can tell when the command has completed when the Get Pose button is no longer disabled. When this is done, the filtered set of particles appears (see Figure 26-12).

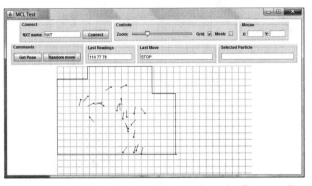

Figure 26-12: Filtering the particles after the first reading

Now it is time for the robot to go out and explore the environment to see if it can figure out which one of these particles in the particle set is closest to its position. Press Random Move and the robot will travel and rotate to a random location. After it is done the random move, wait a moment and you will see the program generate another particle set (see Figure 26-13). This time they are clustered around the remaining particles from the previous particle set.

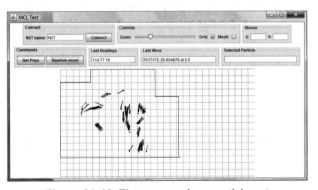

Figure 26-13: The new random particle set

Now press Get Pose again and the robot will take three new range readings from the new location. After a moment, the MCL algorithm will weed out many of the clusters (see Figure 26-14).

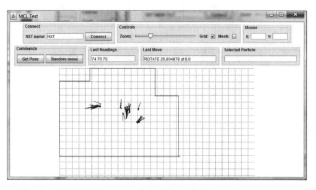

Figure 26-14: Clusters forming after a random move

Alternate pressing Get Pose and Random Move until you are left with just one cluster that represents the true pose of the robot. This means it has pinpointed a coordinate and it will draw a blue circle (see Figure 26-15). If the robot never locates itself, perhaps your test course does not contain any unique features (in other words, it is too symmetrical).

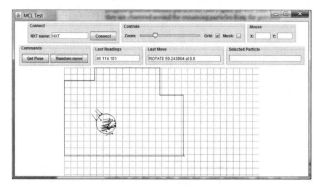

Figure 26-15: A single cluster—the robot is located!

Results

How well does the program work? A lot of the time the program works well and locates the robot within a few centimeters and keeps tracking it as it moves. However, some of the time the program fails. Here are some of the reasons.

- Insufficient particles

  Because of the limitations of NXT memory, the number of particles we can use is limited. Sometimes there is no particle generated with a pose close enough to the pose of the robot. In this case the robot will often localize itself in the wrong location, and eventually will become lost. When this happens we start again with a new set of particles. The best solution is to decrease the size of the map. Another possible solution is to run the particle filtering on the PC. It is relatively easy to change the program so that particle filtering is done on the PC and commands are sent to the NXT to take readings and make moves. However, running the algorithm on the PC limits the applications that it is useful for. For autonomous robots the algorithm needs to run on the NXT.

- Parallel to wall

  When one of the readings taken by the robot is close to parallel to the wall is may get a very different reading than the theoretical one. This can cause the robot to lose track of its location. This is because of the cone effect—the ultrasonic beam is not a thin one-dimensional straight line, which the probability model assumes. A more sophisticated model could be developed to take account of this.

- Too far from walls

  This strategy works well for indoor environments with walls, but if you plunk a robot into the desert, it can wander for miles without encountering any landmarks. Would need to rely on navigation where it uses external landmark (GPS) or the robot drops its own waypoint markers that it can monitor (visually, with a laser, or some other method). The current version of the algorithm ignores readings when any one of the set is undefined because the robot is too far from a wall. When readings are ignored the particles spread out further and the accuracy of localization reduces. If the robot remains far from the wall for a long time, it will get lost. When a reading is made at an acute angle to the wall, you often get an undefined reading even when the wall is in range due to the glancing angle of the ultrasonic sensor. The probability model could be improved to use partial sets of readings.

- Rogue readings

    Sometimes when a wall is near the limit of the ultrasonic sensor's range, a reading that is hugely inaccurate is given. This causes the robot to get lost. The probability model could be changed to ignore large readings and only use the sensor in the range that it is more accurate.

Localization Package

So how does all this work in the leJOS NXJ API? The MCLPoseProvider uses an MCLParticle class to represent a particle, and an MCLParticleSet class to represent the complete set of particles. Both of these classes are includ-

Just for fun!

If you pursued the whole house map from chapter 13, you can show off the abilities of your robot to a friend. With the map on screen, ask your friend to put the robot down anywhere in your home. Don't watch. Stay in your room with the door closed and music playing so you can't see or hear where the robot is placed. When your friend comes back, try a few moves and see if the robot can narrow down where it is. You've done it! For an extra challenge, try extending the example in this chapter so you can take control of the robot and drive it all the way back to your room.

ed in the lejos.robotics.localization package. The important methods in these classes are explained below.

The PC program includes the same lejos.robotics.localisation classes as the NXT program. They are included in pc-comm.jar as well as classes.jar.

We need quite a lot of particles in order to generate one or more that closely match the initial pose of the robot. The memory of the NXT is limited, so there is a limit to how many particles we can use. We have chosen to use 200 particles as the NXT can handle this and it gives a reasonable spread across our mapped area (see the constant NUM_PARTICLES in the PC program). You may need to modify this value if your mapped area is significantly smaller or larger than the one used in this chapter. Note, however, that the NXT cannot support a lot more than 300 particles, so make sure your mapped area is not too big.

In order to explore the environment, the MCLPoseProvider class needs to generate random moves and apply them to the robot and to each particle.

To apply the same move to each particle, the MCLPoseProvider uses the applyMove() method in MCLParticleSet, which uses the applyMove() method in the MCLParticle class. Random noise is added to each move proportional to the distance moved or the angle turned.

After each move, MCLPoseProvider takes a set of new range readings and recalculates the weights for each particle. It uses the calculateWeights() method added to the MCLParticleSet class which uses the calculateWeight() method in the MCLParticle class.

After calculating the weights, it needs to resample the particles. This uses the resample() method in the MCLParticleSet class.

There are methods in the MCLParticleSet class to get the

estimated pose of the robot and the x and y range of the remaining particles, so the accuracy of the estimate can be shown.

MCL on the NXT Only

It is actually possible to run MCLPoseProvider solely on the NXT brick without using your PC. We have a demonstration of this code in the samples directory called Homer. The code is documented. Feel free to modify the code in this sample to your own project requirements.

MCL Theory

That's it for the MCL projects in this chapter. If you are interested in more details about the Monte Carlo algorithms, read on.

Sensor Probability Model

How does the MCL algorithm accurately determine location despite merely having a set of approximate readings? By using a sensor probability model that tells us how probable it is that we would get this set of readings given any particular pose. The probability model takes into account the error in the ultrasonic sensor readings. It follows a normal distribution and assumes that the three readings are independent.

The probability is calculated as follows:

If the robot is at a pose whose distance to the wall in front is x, and we get a reading y, the probability of getting that reading is:

$$\alpha e^{-\frac{d^2}{2\sigma^2}}$$

In case the notation above is not clear, æ is to the power of the fractional part of the equation. Where d is the difference between the actual and theoretical reading (y - x), α is a constant and σ is the standard deviation of the normal distribution. The values of σ is determined by how accurate the sensor is. Sensors with more random noise in their measurements have a larger value of σ.

To get the probability for a set of three readings, we multiply the three readings together. The constant α will be ignored as we will only be interested in the relative probabilities between different poses.

Particle Set and Probability

The probability model takes into account the error in the ultrasonic sensor readings. If we could use our probability model to check the probability of getting a given set of readings against every possible pose of a robot in our mapped area, we could determine the most probable robot pose.

However, there are two problems in doing this. First, it would be very time-consuming to check every pose in the mapped area and calculate the probability. To attempt this we would have to divide the area into small grid squares, and the heading angle into small units, but there would be so many combinations of these that the processing time would be very great. Second, as previously mentioned, there will be many poses with similar probabilities and one set of readings is insufficient to establish the robot pose.

To solve the second problem we need to let the robot explore our mapped area and continue to take measurements. As it applies the probability model to these measurements, this eliminates a lot of the possible poses for the robot and it gradually narrows down the possible poses of the robot. However, to follow all possible initial poses is an impossible task—it would take vastly more processing power than the NXT ARM chip has available to it. This is where particle filtering comes to play.

Rather than tracking all possible starting positions, the algorithm follows the path of a sufficiently large random sample of such initial poses. To do this, the algorithm must generate a set of *particles* with a random pose and make each particle follow the movement of the robot as it randomly explores its environment (see Figure 26-6). After each move, the algorithm uses our calculated probability as a *weight* for each particle. It then *filters* the particles—keeping and duplicating those with the highest weights and replacing those with lower weights. A constant number of particles are maintained throughout. Those particles with the highest weights represent the most likely poses for the robot.

Note that when the robot moves about, even though the lejos.navigation classes use the built-in tachometers to implement very accurate movements, there is still a degree of inaccuracy. A movement probability model is used to estimate this. It is modeled as random noise proportional to the distance moved following a normal distribution. This random noise causes particles that start in the same position to spread out after each random move. You will see this on your PC monitor when we run the program.

The particle filtering algorithm can be used for many problems, not just robot localization.

Further Development with MCL

There are many directions in which you could use the MCL code in the leJOS project:

- Real World Map Data

  To try out a real world example, get out your tape measure and map out an actual room in your house, such as your kitchen or living room, complete with chairs and other obstacles.

- Autonomous robot

  The purpose of the PC program is to give a good graphical representation of the particle filtering algorithm at work. Once you have the algorithm working reliably, you can move the control of the program to the NXT and omit the PC program. You could display the position on the LCD screen.

- Adding obstacles

  You can add obstacles to the map. Adding circular or elliptical obstacles would not be too difficult.

- More sensors or more readings

  If you have multiple Ultrasonic sensors or other range sensors you can speed up the range reading portion of the program. To use multiple ultrasonic sensors, program a new range scanner that implements the RangeScanner interface. Only a couple of lines of the NXT program would need to change. For example, instead of this:

  ```
  RangeScanner sonic = new
  FixedRangeScanner(pilot, sonic);

  scanner.setAngles(ANGLES);
  ```

  Use this:

  ```
  RangeScanner = new new
  MultiRangeScanner(sonic1, sonic2, sonic3);
  ```

The MultiRangeScanner class is the class you program using the RangeScanner interface. The angles in the constructor are set to fixed values, depending on where you point the sensors on the front your robot (probably 45 degrees apart).

Alternatively, if you only have one range ultrasonic range sensor, you could mount the ultrasonic sensor on a turntable and take more than three readings. To use a rotating range scanner, use the class RotatingRangeScanner. Simply use the proper parameters in the constructor as instructed by the Javadocs:

```
RotatingRangeScanner(motor, sonic, gearRatio)
```

CHAPTER 27

GPS

TOPICS IN THIS CHAPTER

- GPS Localization
- Obtaining a GPS
- The Location API
- Plotting Waypoints

So far the projects in this book have been confined in-doors. If you decide to follow the projects in this chapter and the next, your NXT brick will finally be able to go outside! Although the NXT is used primarily for indoor use, you can create LEGO robots that are suitable for the outdoors. The first obstacle to overcome is rough, uneven terrain. Once this barrier has been defeated, you can keep track of coordinates using a GPS receiver, which works great outdoors.

GPS Localization

The Global Positioning System (GPS) is the most widely used navigational system to help travelers get from point A to point B. The system was developed by the United States Department of Defense starting in the 1980s and came online in 1993. Best of all, it was made available for civilian use, which has resulted in a booming commercial market for GPS technology. There are two other satellite navigation systems in the world: GLONASS, a Russian system that fell into disrepair and was restored in 2009, and Galileo by the European Union, which should come online in 2014.

As of this writing there are about 32 satellites in the worldwide GPS constellation orbiting at an altitude of 20,200 kilometers. Because of the shape of the earth, only six to ten satellites may be visible at any given time, with the remainder hidden by the horizon. Each of these satellites sends a signal consisting of the time and precise orbital information, which can be picked up by a *GPS receiver.* The reason they are called receivers is because they merely receive radio signals from satellites. Because the signal takes time to travel from the satellite to a GPS receiver on earth, the distance to the satellite can be calculated using the delay it took for the signal to arrive. In order to achieve a precise and accurate measurement of the delay, atomic clocks are used in the satellites that rely on radioactive decay to obtain the most precise time possible. Once the GPS receiver has obtained the distances to four or more satellites it can identify the location of the GPS receiver.

GPS is a practical solution to the localization problem. It is widely available and easy to use with the LEGO NXT. Most importantly, it has no drift problems. As we saw in earlier chapters, whether using tachometers or a compass for navigation, after a while the robot loses its location estimate. GPS uses fixed landmarks (satellites) to calculate location, eliminating the accumulation of errors we observed before.

Unfortunately, as GPS exists today it is great for outdoor robots but largely impractical for small indoor mobile robots for the following reasons.

1. GPS does not work very well indoors, especially in large buildings (GPS generally works well in suburban houses).

2. GPS has a margin of error of about two to five meters for most GPS units. This is fine for large vehicles and humans operating in wide open spaces over long distances. However, when you are only four inches tall and navigating in an area of few square meters, it does not prove very helpful.

3. Each GPS reading takes one or more seconds for trilateration (using geometry to determine the position). This means the reported location of your robot will lag slightly behind the actual position when your robot is on the move.

For these reasons, we will take the GPS outdoors where the robot can roam across open spaces. This chapter and the next will examine GPS units, demonstrate GPS and how to access coordinates through the location API, and finally build an effective outdoor navigating robot.

Choosing a GPS

There are several options for equipping your NXT with a GPS receiver. The easiest option is to purchase a GPS designed for the NXT brick, called dGPS by Dexter Industries. The other option is to purchase a Bluetooth enabled GPS. Both of these units have differences that might tip your decision in favor of one over the other (see Table 27-1).

	Dexter Industries	Holux Bluetooth
Battery	3V Button Cell	Rechargeable
JSR-179 support	Yes	Yes
Altitude	No	Yes
Waypoint functions	Yes	No
Data bus	I²C Port	BT Wireless
Memory use	Least	Most
Price	$90	$49.95

Table 27-1: Comparing GPS receivers

A Bluetooth GPS can be paired with your NXT brick, but it can also be paired with other devices such as your computer, a handheld computer, or a smart phone. The Holux GPS featured in this chapter also has the ability to produce more data than the Dexter Industries GPS, such as altitude and miscellaneous satellite information. However, the Dexter Industries GPS also has functions that a standard GPS does not, such as calculating distance and angle to a waypoint target. Let's examine each of these receivers.

Dexter Industries

The dGPS is easy to use. Simply plug it into one of the sensor ports, turn on the NXT and it will begin searching for satellites. After about forty seconds, assuming you have an unobstructed view of the horizon, the blue LED light will light up (see Figure 27-1). This means it has acquired the required three satellites and is ready to produce coordinates.

Figure 27-1: Dexter Industries dGPS

The dGPS is fully integrated into the leJOS classes. You can either use the GPSSensor class, located in the lejos.nxt. addon package, or the standard Java Location package (more on this later).

The GPSSensor class has methods not available to standard GPS receivers. For example, you can set a target destination using the methods setLatitude() and setLongitude(). Then you can get the angle and distance to the target at any time using the methods getAngleToDest() and getDistanceToDest().

Website:

For more information, visit:

www.dexterindustries.com

The following simple program allows you to set the latitude and longitude of a target using a button press. Try hiding an object at the location, then pressing the button to record the target. Now that your treasure is hidden, give the NXT brick to someone and see if they can find the hidden treasure.

```
import lejos.nxt.*;
import lejos.nxt.addon.GPSSensor;

public class GeoCache implements
ButtonListener {
  static GPSSensor gps;
  static boolean enabled = false;
  static float start_dist;

  static final int lo = 200;
  static final int hi = 2800;

  public static void main(String[] args)
throws Exception {
    gps = new GPSSensor(SensorPort.S1);
    GeoCache listener = new GeoCache();
    Button.ENTER.addButtonListener(listener);
    Button.RIGHT.addButtonListener(listener);
    Button.ESCAPE.addButtonListener(listener);
    System.out.println("ENTER = set");
    System.out.println("RIGHT = begin");
    System.out.println("ESCAPE = exit");

    while(true) {
      if(enabled) {
        float distance = gps.getDistanceToDest();
        float percent = (distance/start_dist);
        int freq = (int)(lo + (hi - (percent * hi)));
        System.out.println("Dist:" + distance);
        Sound.playTone(freq, 200);
      }
      Thread.sleep(1100);
    }
  }
  public void buttonPressed(Button b) {}

  public void buttonReleased(Button b) {
    switch(b.getId()) {
    case Button.ID_ENTER:
      gps.setLatitude(gps.getLatitude());
      gps.setLongitude(gps.getLongitude());
      System.out.println("Target set!");
      break;
    case Button.ID_RIGHT:
      start_dist = gps.getDistanceToDest();
      enabled = !enabled;
      break;
    case Button.ID_ESCAPE:
      System.exit(0);
    }
  }
}
```

This game is really fun for kids and adults. It is basically a version of Geocaching, except more spontaneous. It is basically hide and seek using objects instead of people and a GPS. One person hides an object somewhere in the neighborhood, and another person tries to find it using the GPS.

1. To play spontaneous Geocaching:

2. Plug the dGPS sensor into port 1. Turn on the NXT.

3. Walk somewhere in the neighborhood and drop an item on the ground.

4. Press the Enter button to set the target location.

5. Walk back home and give the NXT to a friend.

6. Your friend can press the right-arrow button on the NXT to start and then begin tracking the hidden object.

Bluetooth GPS

GPS receivers use the NMEA 0183 standard to serve data, so any Bluetooth GPS unit should work fine with leJOS NXJ, no matter what manufacturer you choose. If you have not yet purchased a Bluetooth GPS receiver, there are several factors to consider. Ideally you'll want something small, cheap, rechargeable and with good reception (indoors and outdoors).

As of this writing, Magellan and Garman are the most popular manufacturers. Many of their models incorporate a Bluetooth data connection, which is compatible with the leJOS NXJ API. However, they are a little on the large size and you'll pay a premium for the convenience of an LCD screen.

Another company named Holux makes a line of wonderful Bluetooth GPS receivers that suit our purposes, including the Holux-1200 (see Figure 27-2) or the Holux GPSlim 240. Both are the size of a lighter, which makes it perfect for attaching to a small robot. It receives satellite radio signals well, even indoors, using an incredible microstrip patch antenna (an antenna printed on the circuit board). It is re markably accurate (see the test with Google maps below). The battery lasts for about eight hours before it needs another recharging via the USB port. And it is relatively cheap, especially on eBay where it can be obtained for under $50.

Figure 27-2: The Holux-1200

Once you turn on your GPS receiver it takes a while to "warm up." The Holux receivers start acquiring satellite signals quite fast after being turned on compared to others on the market. It takes roughly 40 seconds to acquire the many satellite signals and download the orbital information from all the satellites. Once that is completed it can begin calculating coordinates.

Once the receiver has acquired enough satellites it begins outputting NMEA sentences via Bluetooth. The Holux receiver produces five different NMEA sentences: GGA, GSA, GSV, RMC, and VTG. It produces these sentences in order, constantly looping every 950 ms seconds. Because of this cycle, you can count on location data being no more than a second old when you read it, which is close to real time.

Just for fun!

Programs can interact with GPS receivers using a variety of protocols, but the standard protocol supported by almost all GPS receivers is NMEA. This protocol was developed by the National Marine Electronics Association for use with marine devices, such as navigation instruments, depth finders, and of course GPS receivers. When a GPS is set to NMEA mode, it continuously outputs data from the GPS unit. A typical sentence is shown below:

$GPGLL,4951.3637,N,09706.1 689,W,185113.203,A*2E

Holux and Navibe both make affordable, portable Bluetooth GPS units that are widely available on eBay:

www.holux.com

www.navibe.com

The Location API

If you are excited about using GPS in your projects then you will find a lot of good things in leJOS NXJ, which includes the official Java Location API, otherwise known as JSR-179. Although this API was written primarily for the mobile phone market, it is still very useful and functional with the NXT brick.

You can obtain a lot of useful data from the location classes, located in the javax.microedition.location package. The ultimate goal is to get various location data from the Location and Coordinates classes. The Coordinates class is retrieved from Location, which in turn is retrieved from a LocationProvider object (see Figure 27.3). But before you can do any of this, you need to establish a connection with a GPS unit. First, let's deal with the Bluetooth GPS unit, and then the dGPS unit.

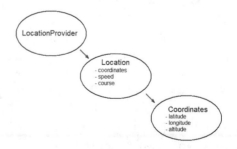

Figure 27-3: Obtaining Coordinates

Theoretically the LocationProvider allows you to choose which Bluetooth GPS receiver to connect to. In a classroom or other crowded setting, there could be more than one receiver, so it would be beneficial to specify which one to con-

nect to. However, the official API is not very good at allowing the programmer to choose the GPS unit. Instead, it uses
a strange method to determine which GPS to connect to by
using a Criteria object, which only has general criteria for
things like power consumption, accuracy, and cost. None of
these are really helpful with the NXT, so instead of supplying
a Criteria object you can just supply null, which will cause
the LocationProvider to connect to the least restrictive device.

```
LocationProvider lp = LocationProvider.
getInstance(null);
```

Now the obvious question: How do you actually choose
which GPS to connect to? As a programmer, you can't. Only
the user can. The NXT will connect to the first *paired* GPS
receiver it finds via Bluetooth. So it is up to you to choose
the proper GPS receiver by pairing it at the menu. Once
paired, make sure it is turned on and it will connect. In a
classroom setting with multiple GPS receivers, each one
should be paired with an NXT brick to avoid confusion. That
way the NXT brick won't connect to someone else's.

If you are using a dGPS receiver, you need to use a special
type of Criteria object. The difference with the dGPS is that
it must specify a port. This is done as follows:

```
dGPSCriteria criteria = new
gGPSCriteria(SensorPort.S1);
LocationProvider lp = LocationProvider.
getInstance(criteria);
```

Once the LocationProvider is instantiated it will connect automatically and then you can get location information using
the getLocation() method. You can also provide a timeout
value so that if it can't retrieve a location within the specified time it will return nothing. In practice, once the GPS is
warmed up it will provide data under any circumstances,
since it gives the last one received. For this reason I recommend using -1, which uses the default timeout:

```
Location l = lp.getLocation(-1);
```

The Location object is a container with a variety of useful location information. The most important data in Location is the Coordinates object, which contains latitude, longitude, and sometimes altitude. In Chapter 8 we learned about Cartesian coordinates, which consisted of an x and y position. The Location API uses a similar set of coordinates, except they are not relative to the arbitrary starting location of the robot. Instead, these coordinates are mapped onto a sphere—the earth. The latitude and longitude readings are in degrees. Latitude ranges from -90 degrees at the South Pole to +90 degrees at the North Pole (0 degrees at the equator). Longitude starts at 0 degrees at the Prime Meridian and increases eastward up to +180 degrees, while westward it decreases to -180 degrees (see Figure 27-4).

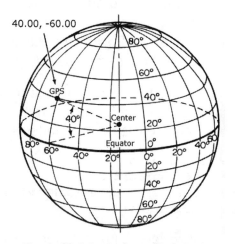

Figure 27-4: Latitude and Longitude

Traditionally, these values have been reported in degrees, minutes, and seconds (see Table 27-1). You might be asking why we are suddenly talking about time? Minutes and seconds don't really apply to degrees, do they? Well, it's an old system that was developed when mariners used clocks to estimate longitude and is no longer really relevant in this age, although some people still use them.

The traditional notation is reported in the form DDD MM SS.SS. As stated above, the degrees values (DDD) range from 0 to 180 for longitude and 0 to 90 for latitude. Minutes, on the other hand, range from 0 to 59. Seconds range from 0 to 59.99 (the number decimal places for seconds is not fixed).

This is an archaic system when it comes to computers, which use plain decimal numbers. Therefore, most applications use *decimal degrees*. This is just a degree value ranging from -180 to +180 for longitude, and -90 to +90 for latitude (see Table 27-2). East and North are positive numbers, while West and South are negative numbers.

	Longitude	Latitude
Format	DDD MM SS.SS	DD MM SS.SS
Traditional example	97° 05' 52.5402" W	49° 49' 10.57" N
Decimal degrees	-97.097928	49.819603

Table 27-2: Converting to decimal degrees

Using the location package, you can also retrieve other information. The Location class provides speed, course (heading) and the timestamp, which will be demonstrated shortly.

There is also a class called QualifiedCoordinates which produces the horizontal and vertical error, in meters. This margin of error is calculated using the number of satellites and their position. Use getHorizontalAccuracy() and getVerticalAccuracy() to retrieve these values.

To make navigation updates easier, you can retrieve Location data at intervals by using a LocationListener. The listener can be notified in intervals of one second (not fractions of a second) or it can be notified exactly when a new location becomes available. The latter is very useful if you want your robot to operate with the freshest data possible.

There are several methods in the package which are outstanding for navigating. These methods calculate metrics for distance and direction (azimuth) to another coordinate. These complex calculations even take into account the curvature of the earth. Both methods are called from the Coordinates class, as follows:

```
Coordinates.azimuthTo(Coordinates)
Coordinates.distance(Coordinates)
```

The final piece of the location API we will examine is a system for navigating from one waypoint to another. This is accomplished with the ProximityListener, which is notified as soon as the GPS receiver enters within a specified proximity of a set of coordinates.

```
LocationProvider.
addProximityListener(
ProximityListener pl,
Coordinates c,
float proximityRadius)
```

Now that you've seen an overview of these classes, we will try two different projects. The first merely displays and records the current location, while the second project attempts to navigate through a series of waypoints.

Just for fun!

You can convert decimal degrees to traditional latitude/longitude values at the following website:

http://transition.fcc.gov/mb/audio/bickel/DDDMMSS-decimal.html

Plotting Waypoints

In this section we will develop a program to display latitude and longitude. It will also save the coordinates to the NXT file system each time you press the orange button (this feature will be used in the next chapter for the navigator robot).

The code below is straightforward. The main() method obtains an instance of LocationProvider. It then creates a LocationListener and adds it to the LocationProvider using the default parameter -1 for the interval so it is notified when a new location is available. The listener method locationUpdated() merely redraws the coordinates to the LCD screen. The remainder of the code (including the loop) writes the coordinates to the properties object, and then when the exit button is pressed it writes the properties to the file system.

Just for fun!

The leJOS API contains several classes that are not very useful for small LEGO-sized robots, including AddressInfo, Landmark, and LandmarkStore. AddressInfo contains street information, which can presumably be used with Google Maps and other services to find a route from one location to another. (You can also use Google Moon or Google Mars if you ever succeed in launching your LEGO robot onto another celestial body.)

The DARPA Urban Challenge in 2007 proved that autonomous robots can navigate roads and intersections. Using the JSR-179 API, you have all the necessary classes to navigate city streets from one address to another—other than the fact that your LEGO vehicle would probably get crushed under traffic (and government regulations) you could probably construct your own automated taxi service.

```java
import lejos.nxt.*;
import javax.microedition.location.*;
import java.io.*;
import java.util.Properties;

public class WaypointPlotter implements
LocationListener {

  public static void main(String[] args) {

    // Get an instance of the provider
    LocationProvider lp = null;
    try {
      System.out.println("Connecting..");
      lp = LocationProvider.getInstance(null);
    } catch (LocationException e) {
      System.err.println(e.getMessage());
      Button.waitForPress();
      System.exit(0);
    }

    LocationListener locationListener = new
WaypointPlotter();
    lp.setLocationListener(locationListener,
-1, 6, 0);

    Properties p = new Properties();
    Button.setKeyClickTone(Button.ENTER.
getId(), 0); // Disable default sound
    int count = 0;
    int press;
    do {
      press = Button.waitForPress();
      if(press == Button.ID_ENTER) {
        Sound.beepSequenceUp();
        Location l;
        try {
          l = lp.getLocation(-1);
        } catch (Exception e) {
          System.err.println(e.getMessage());
          continue;
        }
        Coordinates c =
l.getQualifiedCoordinates();
        p.setProperty("lat" + count, "" +
c.getLatitude());
        p.setProperty("long" + count++, "" +
c.getLongitude());
      }
    } while(press != Button.ID_ESCAPE);
```

```
      p.setProperty("waypoints", "" + count);
      File f = new File("waypoints.prop");
      if(!f.exists())
        try {
          f.createNewFile();
          FileOutputStream out = new
  FileOutputStream(f);
          p.store(out, "GPS Waypoints");
          out.close();
        } catch (IOException e) {
          System.err.println("Save failed");
          System.err.println(e.getMessage());
          Button.waitForPress();
          System.exit(0);
        }
      }

    public void locationUpdated(LocationProvid
  er lp, Location l) {
        QualifiedCoordinates c =
  l.getQualifiedCoordinates();
        LCD.clearDisplay();
        LCD.drawString("Lat " + c.getLatitude(),
  0, 1);
        LCD.drawString("Long " +
  c.getLongitude(), 0, 2);
        LCD.drawString("Hor.Acc " +
  c.getHorizontalAccuracy(), 0, 3);
        LCD.refresh();
      }

    public void providerStateChanged(LocationP
  rovider lp, int state) {}
  }
```

NOTE: Owners of the dGPS should substitute this line:

```
lp = LocationProvider.getInstance(null);
```

With the following lines of code:

```
dGPSCriteria crit = new
gGPSCriteria(SensorPort.S1);
lp = LocationProvider.getInstance(crit);
```

To test the program, turn on your GPS receiver and NXT brick. Wait until the blue light on the dGPS is lit up before running the program.

Just for fun!

When the United States Department of Defense conceived of GPS, they hardly thought civilians would use it as a platform for games. In fact there are many interesting games, and more being invented every day. The most popular GPS game is called Geocaching, a form of treasure hunting. Visit the website below and enter your ZIP code to find hidden treasure ('caches') in your area. You can alter the WaypointPlotter code to display the distance and azimuth to a Geocache destination. You can also use the Coordinates.convert() method to convert decimal degrees to degrees and minutes (the format used by Geocachers).

www.geocaching.com

If you are using the Holux-1200, it has two LED icons. The orange satellite icon remains solid until it has acquired enough satellites to create a fix, however you can connect using Bluetooth at anytime. Under the Bluetooth menu, search and pair your GPS (the PIN is likely 0000). When connected, the Bluetooth icon goes from blinking every half second to blinking every second. Once that is done, compile and upload the code to your NXT brick, then run it. The program will connect to the GPS and begin displaying your present coordinates.

Troubleshooting

There might be several reasons why the program isn't connecting to your GPS unit, usually apparent by the displayed error message. In most cases, either you haven't paired your GPS yet at the main NXT Bluetooth menu, the PIN number is wrong (again, at the menu), or your GPS unit is not turned on. You might rarely encounter too much local radio interference and Bluetooth did not connect. In these cases, just try again.

My Bluetooth GPS was working for months, and then one day it stopped working with my NXT for no apparent reason. The solution was to remove the device from the Bluetooth menu, then find and pair again.

Just for fun!

Once you have your program running and coordinates displayed, try going to Google Maps:

http://maps.google.com

You can type in the latitude and longitude values (separated by a comma). When you hit enter it will show the location (see Figure 27-5). Click on the satellite imagery and then zoom in as far as possible to see how close it is to your actual location. It's amazing how accurate the location is displayed. My position showed up within a few meters of my actual location in the south-easterly part of the house.

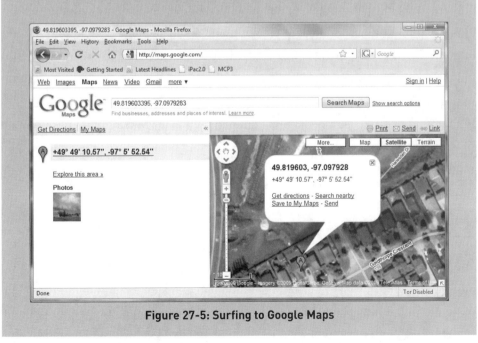

Figure 27-5: Surfing to Google Maps

CHAPTER 28

GPS Waypoint Following

TOPICS IN THIS CHAPTER

- Outdoor Robots
- Building the Grass Master
- Programming the robot
- Using the robot

In the previous chapter we programmed the NXT to record GPS waypoints and save them to a file. This chapter is the second half of that project, in which a robot will traverse those waypoints using the onboard GPS receiver.

The Grass Master

The dream of autonomous vehicle navigation has been alive for a long time. In 2005 this dream was realized by the DARPA Grand Challenge, in which teams from around the world competed on an off-road course filled with natural obstacles and hazards. Of the 23 finalists, five managed to complete the 212 kilometer course.

In this section we will build a small scale version of a DARPA vehicle called Grass Master (see Figure 28-1). It won't attempt to avoid obstacles and hazards, but it will follow a course over semi-rough terrain (an open field with short grass).

Figure 28-1: The Grass Master preparing to conquer the outdoors

I was a little worried the small tires in the 1.0 and 2.0 kits would be unable to handle outdoor terrain, but the Grass Master robot handles it admirably, even over small mounds and depressions. With the balloon tires of the NXT 1.0 kit the robot really tears through the grass at a good pace. The smaller street racing tires in the 2.0 kit cause a speed decrease but the robot is still sure footed. The three-point wheel base, much like a futuristic tractor, ensures that all tires are propelling the vehicle forward at all times. The only downside was that my black tires turned a little brown due to dust and grime.

The Grass Master uses 100% LEGO NXT parts (1.0 or 2.0). While other robots are very versatile, able to move forward and backward, and steer well in any environment, the Grass Master robot is not versatile. It is made for one thing: driving forward over grass. It can't turn in one spot and must be moving to initiate sweeping turns. When making a hard turn, the turning radius is about 60 centimeters (two feet) on grass, and much wider on carpet.

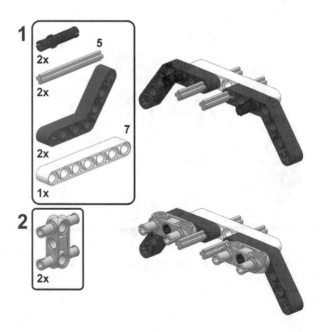

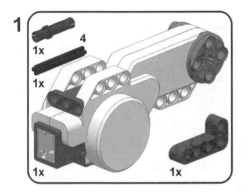

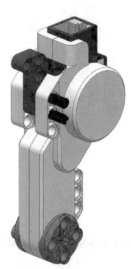

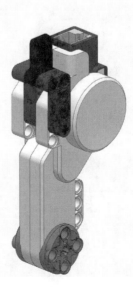

3

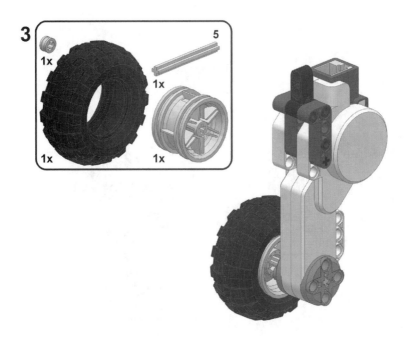

1

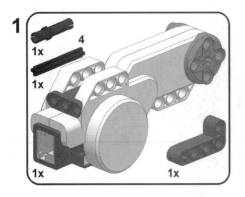

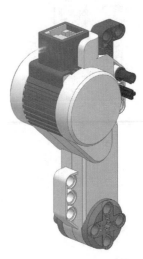

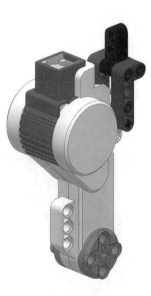

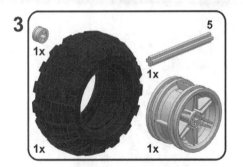

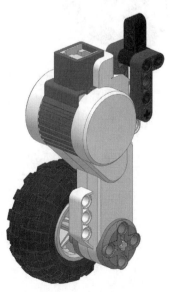

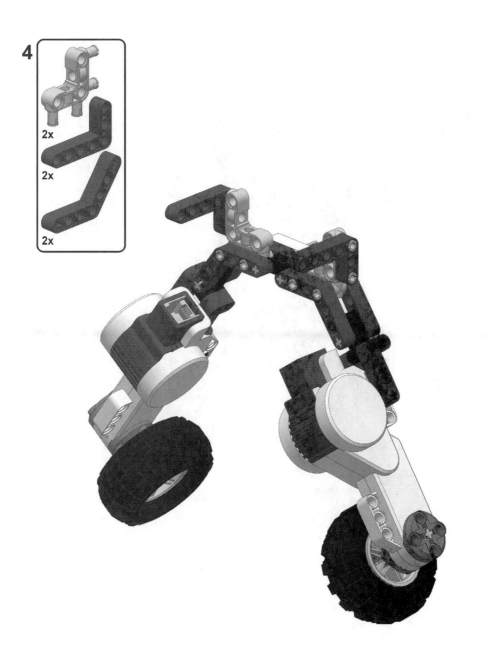

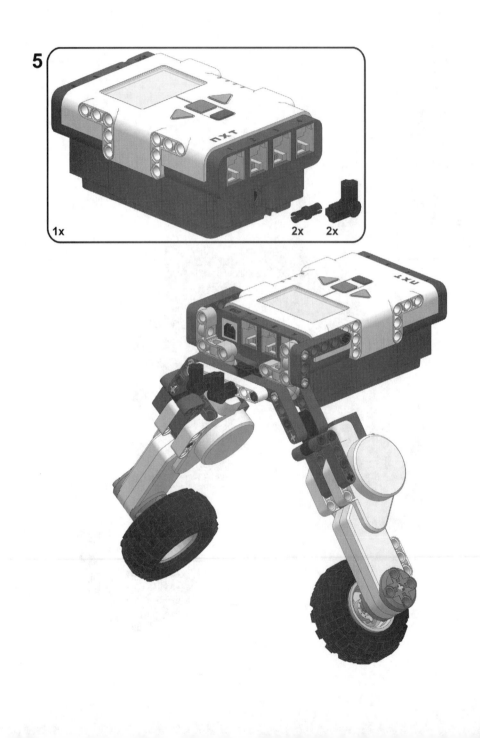

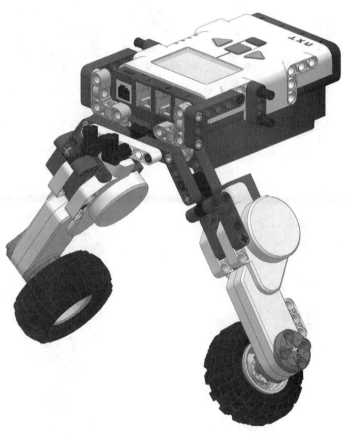

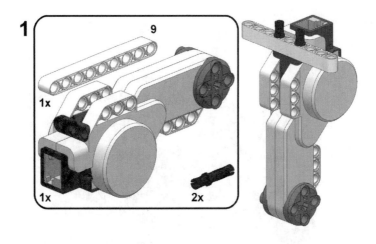

1

9

1x

1x

2x

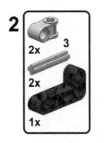

2

2x

3

2x

1x

3

2x

2x

1x

7

2x

4

2x

2x

7

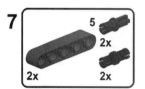

Cables

Now plug in the cables for the motors. Use a short cable to connect the center motor to port A. Use medium cables to connect port B and C to the right and left motors respectively. A small GPS receiver, such as the Holux-1200, fits easily on the back of the robot, pinned down by the wires (see Figure 28-2). I tried to come up with a 100% LEGO solution to hold the GPS receiver, but if you are concerned about losing your costly item in the grass, try using a rubber band to secure it to the robot.

Figure 28-2: Securing the GPS Receiver

Coding a Waypoint Follower

The Grass Master robot has the simple task of driving from one waypoint to another. To complete the task, it starts driving forward and does not stop until it reaches the last waypoint. The code will attempt to steer towards a target by measuring the current heading, calculating the azimuth to the target, and correcting the steering accordingly.

During its travels, the GPS updates the program about once every second with a new set of coordinates. However, if the robot is driving forward and just needs to continue forward, it shouldn't stop and then start again. It should just keep driving forward. So to avoid starting and stopping, the robot will have various states such as steer left, steer right, and drive straight. This will prevent the robot from reacting to every event, causing jerky start and stop motion.

```
import lejos.nxt.*;
import javax.microedition.location.*;
import java.io.*;
import java.util.Properties;

public class WaypointFollower implements
ProximityListener {

  public static final int MAX_SPEED = 900;
  public static final int MED_SPEED = 700;
  public static final int MIN_SPEED = 500;

  private static boolean waypointReached = false;

  public static void main(String[] args) {
    powerSteer(MAX_SPEED, MAX_SPEED);
    WaypointFollower proximityListener = new
WaypointFollower();

    // Get an instance of the provider:
    LocationProvider lp = null;
    try {
      System.err.println("Connecting.. ");
    lp = LocationProvider.getInstance(null);
      System.err.println("CONNECTED");
    } catch(LocationException e) {
      System.err.println(e.getMessage());
```

```java
    Button.waitForPress();
    System.exit(0);
  }

  // Load properties:
  Properties p = new Properties();
  File f = new File("waypoints.prop");
  if(f.exists()) {
    FileInputStream in;
    try {
      in = new FileInputStream(f);
      p.load(in);
      in.close();
    } catch (Exception e) {
      System.err.println("Failed to load
waypoints");
      System.err.println(e.getMessage());
      Button.waitForPress();
      System.exit(0);
    }
  } else {
    System.out.println("No waypoints found");
    while(!Button.ESCAPE.isPressed())
{Thread.yield();}
    System.exit(0);
  }

  int waypointCount = Integer.parseInt(p.
getProperty("waypoints"));
  for(int i=0;i<waypointCount;i++) {
    waypointReached = false;
    double latitude = Double.parseDouble(p.
getProperty("lat" + i));
    double longitude = Double.parseDouble(p.
getProperty("long" + i));
    Coordinates waypoint = new
Coordinates(latitude, longitude);

    try {

LocationProvider.addProximityListener(proximi
tyListener, waypoint, 3);
    } catch (LocationException e) {
      System.out.println(e.getMessage());
      while(!Button.ESCAPE.isPressed())
{Thread.yield();}
      System.exit(0);
    }
```

```
while(!waypointReached) {
  Location l;
  try {
    l = lp.getLocation(-1);
  } catch (Exception e) {
    System.err.println(e.getMessage());
    continue;
  }

  Coordinates current =
l.getQualifiedCoordinates();
  double az = current.azimuthTo(waypoint);
  LCD.clearDisplay();
  LCD.drawString("LAT:" + current.
getLatitude(), 0, 0);
  LCD.drawString("LONG:" + current.
getLongitude(), 0, 1);
  LCD.drawString("TLAT:" + latitude, 0, 2);
  LCD.drawString("TLONG:" + longitude, 0, 3);
  LCD.drawString("AZ:" + az, 0, 4);
```

Just for fun!

Location based games are invented every day. Try inventing your own GPS game! I tried making a game similar to Bocce by taping my GPS receiver inside a hamster ball. One person stands at each end of the pitch. The goal is to roll the ball closest to a flag at each end. The NXT monitors the ball to see when it is thrown and when it has stopped. It then calculates how close each person gets to the flag. Players keep throwing the ball back and forth, accumulating error each time. The loser is the first person to accumulate over 50 feet worth of error. Using beeps, players know when they can throw and when the ball has stopped moving.

For more GPS games, visit Wikipedia:

http://en.wikipedia.org/wiki/Location-based_game

```
        double course = l.getCourse();
        LCD.drawString("COURSE:" + course, 0, 5);

        double diff = az - course;
        if (diff > 180) diff -= 360;
          if (diff < -180) diff += 360;
          LCD.drawString("DIFF:" + diff, 0, 6);
          LCD.drawString("DIST:" + current.
distance(waypoint), 0, 7);
          LCD.refresh();

          if(diff < -90) {
            powerSteer(MIN_SPEED, MAX_SPEED);
          } else if(diff < -10) {
            powerSteer(MED_SPEED, MAX_SPEED);
          } else if(diff > 90) {
            powerSteer(MAX_SPEED, MIN_SPEED);
          } else if(diff > 10) {
            powerSteer(MAX_SPEED, MED_SPEED);
          } else {
            powerSteer(MAX_SPEED, MAX_SPEED);
          }
        try {Thread.sleep(500);}
        catch(InterruptedException e){}
      }
    }
  }

  public static void powerSteer(int left, int
right) {
    Motor.A.setSpeed((left + right)/2);
    Motor.B.setSpeed(right);
    Motor.C.setSpeed(left);

    Motor.A.backward();
    Motor.B.forward();
    Motor.C.forward();
  }

  public void monitoringStateChanged(boolean i) {}

  public void proximityEvent(Coordinates c,
Location l) {
    Sound.beepSequenceUp();
    waypointReached = true;
  }
}
```

NOTE: Owners of the dGPS should substitute this line:

```
lp = LocationProvider.getInstance(null);
```

With the following lines of code:

```
dGPSCriteria crit = new
gGPSCriteria(SensorPort.S1);
lp = LocationProvider.getInstance(crit);
```

You'll notice we didn't hard-code the waypoints. There are a few good reasons for saving the coordinates in a properties file rather than hard-coding the data, but the most important reason is so you can use the waypoint tracker at different locations without having to reprogram waypoints in the code. It's also easier to walk around and set waypoints with the push of a button (a one step process), rather than writing them down and entering them into your code (a tedious two step process).

Find an open location with reasonably short grass, such as a football field, golf course or city park (pavement will not work). Before running this program, you need to lay down a course for your robot to follow using the WaypointPlotter code from the previous chapter. Turn on your GPS early and give it at least a few minutes to acquire as many satellites as possible for a stable GPS solution.

While this is going on, walk around the park and place markers for the course. I bought some wire marker flags from a local hardware store (see Figure 28-3). While plotting the course, it was easy to hold the NXT in one hand, and with the other toss the Bluetooth GPS unit into the grass where you want a waypoint, plant a flag next to the GPS, then press the orange button. Watch the GPS coordinates and don't hit the orange button until they have stabilized. Make sure to walk a path that avoids ruts and obstacles. When you have your final waypoint entered, click the grey button to save the coordinates to the properties file. In the file system you will see a file called *waypoints.prop*.

With balloon tires the Grass Master takes about one minute to travel 18 meters (60 feet), while the street racing tires travel 14 meters (46 feet) in a minute.

> **NOTE:** The orientation of the GPS receiver on your robot doesn't matter. It contains no compass sensor to identify direction. It is merely looking at the previous location and comparing it to the current location to identify the present course.

Figure 28-3: Wire marker flag in a field

The vehicle does fine when making gradual course corrections, but sometimes it performs a circle or two before setting off in the right direction. This is because the GPS receiver determines course by comparing the current coordinates to the previous coordinates. If you've ever watched the coordinates while standing still, you'll notice they jump around a lot, even though you are not moving. This is because satellite based location is only accurate to about four meters. When the robot is making a turn and not on a definite forward course, it mixes up the current course. GPS units are primarily good for travelling in cars, where the error margin is wiped out because the car makes so much headway between readings. When stationary, however, the course determination isn't as good. However, it will still eventually sort it out and get to the next waypoint.

There are several problems that can hamper GPS accuracy. On cloudy days, GPS just doesn't seem to be as accurate, probably because it can't receive as many satellites as on clear days. I also encountered more accurate readings in open areas compared to city locations near houses and buildings, where many of the satellites are blocked.

So how can you use this type of robot? Transportation mainly—at least that's why the United States Department of Defense funded the DARPA Grand Challenge. They want one-third of their land vehicles to be autonomous by 2015. In the private sector, this type of robot could be useful for transporting items/people and making deliveries.

Advanced GPS Programming

JSR-179 contains enough classes and methods to perform navigation. However, a Bluetooth GPS receiver supplies a lot more information than JSR-179 allows you to access. If you want to look at detailed data, check out the classes in the lejos.addon.gps package.

The primary class is this package is the GPS class. It contains methods for retrieving the number of satellites, satellite information, error ranges, and current date and time. The leJOS NXJ package comes with a sample application that demonstrates how to connect to the GPS class and retrieve data. It is located in the samples directory GPSInfo.

Accuracy is dependent on the number of satellites and the positions of the satellites in the sky. If they are bunched close together at any given moment, accuracy is decreased. This accuracy is represented by Dilution of Precision values (DOP). There are Vertical, Horizontal, and Positional (3D) DOP values, known as HDOP, VDOP, and PDOP.

Website:

Wikipedia has an excellent article describing DOP:

http://en.wikipedia.org/wiki/Dilution_of_precision_(GPS)

GPSInfo is a useful utility for discovering how well your GPS functions in different locations. Try it indoors and you might see less satellites and lower accuracy values. Outdoors you get lots of satellites, but around buildings you might lose and gain satellites as you move, which constantly changes the

GPS solution. If the GPS solution changes, you will probably see inconsistent waypoints. Open fields offer the most stable and accurate GPS solution. If you live in a mountainous area, the horizon is more obstructed and you will receive fewer satellites compared to flat areas.

Just for fun!

Each satellite in the GPS network has a unique SVN and PRN number. You can use the GPS class to find out which satellites you are receiving by using the getSatellite() method, then use Satellite.getPRN() to return the PRN number. Once you have this number you can look it up online to find out more information on the satellites in your solution, including their launch date and even a photograph. You'll never find SVN 07 because it failed while in orbit, nor will you find SVN 42 because it was destroyed after liftoff. A full listing of GPS satellites is available at:

http://en.wikipedia.org/wiki/List_of_GPS_satellite_launches

Some of the satellite information in GPSInfo is interesting. Azimuth and elevation let you point at the satellite at any time. Face North and rotate yourself 'azimuth' degrees clockwise. Now point your finger at the horizon and raise it 'elevation' degrees. You are now pointing at the satellite. You could even build a robot to point out satellites automatically.

The signal to noise ratio indicates how well you are receiving individual satellites. Using the azimuth and elevation, you can get a good indication of why a satellite isn't being received well, such as if it is behind a building or low on the horizon. The hour of the day also seems to affect the signal to noise ratio. Also, if you run GPSInfo in the morning and then try it 12 hours later, the satellites in orbit over your location will have changed dramatically (use the PRN numbers to identify individual satellites).

Just for fun!

The Coordinates class in the Location API has no problems crossing the International Date Line or other geographical zones. This means these classes will work fine for autonomous land-based treks or even a UAV (Unmanned Aerial Vehicle) as shown in Figure 28-4. If you ever succeed in adapting an RC model airplane control system to the NXT, the Location API will not let you down.

Figure 28-4: A small military UAV

CHAPTER 29

Open Source Development

TOPICS IN THIS CHAPTER

- leJOS Project
- Sourceforge
- Subversion

The leJOS project is over a decade old and has had dozens of developers over the years. It's ever growing and there are always improvements to be made. Now that you are familiar with leJOS, you might have seen a few things that you would like to improve. If so, you might consider adding your own code to the project.

Working on a project like leJOS is a great learning experience. Nothing hones your thinking skills more than spotting a problem and working through all the challenges. There are other benefits as well. Working for an open source project like leJOS can help build your resume. You'll end up with a few things in your portfolio to show employers what you can accomplish. It's also a lot of fun to work with people from around the world who share an intense passion for robots and Java.

The purpose of this chapter is to put a spotlight on the process of open source development. We'll show you how to get a copy of the latest leJOS code on your computer using Eclipse, and how to submit new code to the project. First, let's examine some aspects of the leJOS project.

leJOS Project

If you have an interest in robotics or computer programming, there is probably something in the leJOS project for you. Let's briefly examine some areas of development within the leJOS project.

- Utilities—leJOS has a number of utilities in the bin directory, such as NXJ Control Center. If you have some GUI skills or would like to develop some, there are opportunities to improve these utilities or add your own.

- Web Developer—the leJOS homepage complements the leJOS software. It hosts a forum, tutorials, and acts as the main showcase for our project. There are opportunities to take the webpage even further, such as allowing users to upload instructions for building and programming LEGO robots.

- Video production—the leJOS project would like to host Youtube videos. If you have a talent for filming and editing videos, there are opportunities to develop material for our website.

- Graphics—if you have an artistic streak, leJOS could use your skills developing project logos, website graphics, icons for utilities and buttons for the Eclipse plug-in.

- Documentation—we can always use someone with writing skills to improve our Javadocs and online tutorials. Documenting software is a great way to become thoroughly acquainted with the leJOS API.

- Robotics Packages—the packages that start with lejos. robotics contain classes dealing with specific robotics problems. These include navigation but there are several areas of research that could be added to the project. Have an idea for a neural network package? How about inverse kinematics? If you have some useful classes for cutting edge robotics problems, you might consider adapting them to the leJOS project.

- Java platform—the leJOS project has many of the standard Java classes, but not all of them. Perhaps you require a particular class for your project but it is not available in the leJOS API. In this case, you could implement this class yourself and submit the code to the leJOS project. If the class is useful to you, chances are others will find it useful too.

- Mobile Devices—smartphones and tablet devices compliment the leJOS project well. They allow hand-held control of the NXT brick. There are options to further develop Android and JME support.

- JVM—the leJOS JVM is constantly undergoing refinement. This is the most difficult aspect of development, requiring proficiency in C language and knowledge of the LEGO NXT hardware.

- Porting leJOS—if you spot another device that would be suitable for a small JVM, such as a handheld gaming device, you could try porting the leJOS JVM to this device. This would allow users to program Java games and applications for that device.

Sourceforge Code Hosting

All of the leJOS code is hosted on Sourceforge. Beyond just hosting code on its servers, Sourceforge offers tools to help developers maintain their projects. These tools help facilitate discussion, keep track of to-do lists, host our website, operate forums, keep track of bugs and even host our own developer Wiki.

You will need to use a Sourceforge ID and password if you want to download project code below. If you like the concept of open source software, you might consider signing up. It is easy, just click on register at the following website:

www.sourceforge.net

Installing Subversion Software

If you want to modify the leJOS code, you will need a local copy of the classes on your computer. To keep this code synchronized and up to date with the project code stored on Sourceforge, we use a version control system called Subversion (SVN) hosted by the Sourceforge servers.

Subversion allows you to obtain a copy of the code, check for conflicts between your local code and the server code (in case two people modify the same piece of code at the same time) and submit code to the server.

To use this system, you will need a plug-in for Eclipse called Subversive. Let's install the plug-in.

1. Run Eclipse. Select Help > Install New Software...

2. In the install window, where it says "type or select a site," click the down arrow. From the list, select the update site for your particular version of Eclipse. For example, if you are using Indigo, select Indigo from the list (see Figure 29-1).

3. A list of software categories will appear. Click the arrow next to Collaboration and scroll down to Subversive SVN Team Provider (see Figure 29-2). Place a checkmark next to this software and click Next.

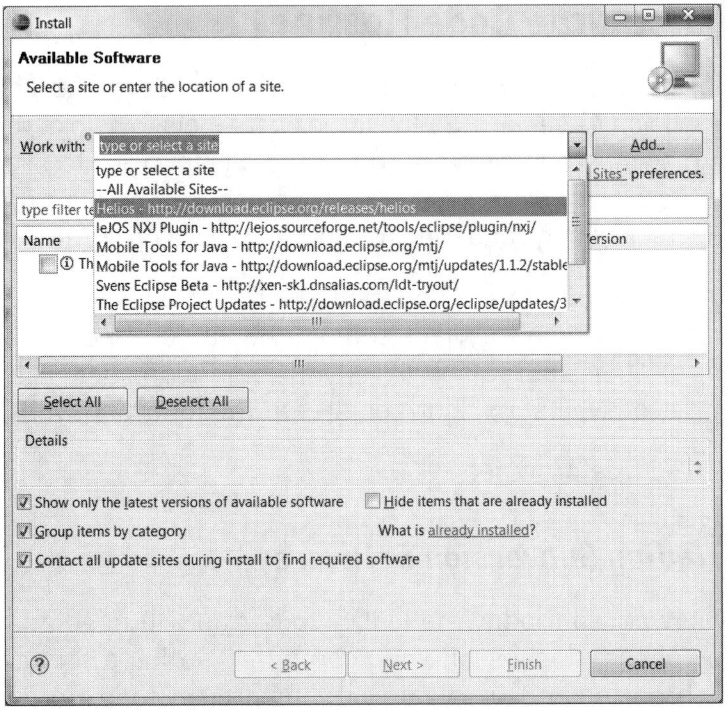

Figure 29-1: Finding official Eclipse plug-ins

4. SVN will calculate requirements and dependencies. When it is done, click Next again. Read the agreement and then select the appropriate radio button. Click Finish.

5. When installation completes, select Restart Now.

6. The introduction page will display the option to view the Subversive Overview. If you click this you will see a list of help topics for Subversive (see Figure 29-3).

7. To go straight into Eclipse, click the workbench icon on the introduction page.

8. Select Window > Open Perspective > Other... From the Open Perspective window, select SVN Repository Exploring (see Figure 29-4). Click OK.

9. There is an extra step here due to licensing restrictions that prohibit including a module in the Subversive download. Select the latest SVN Kit by Polarion (see Figure 29-5). Click Finish.

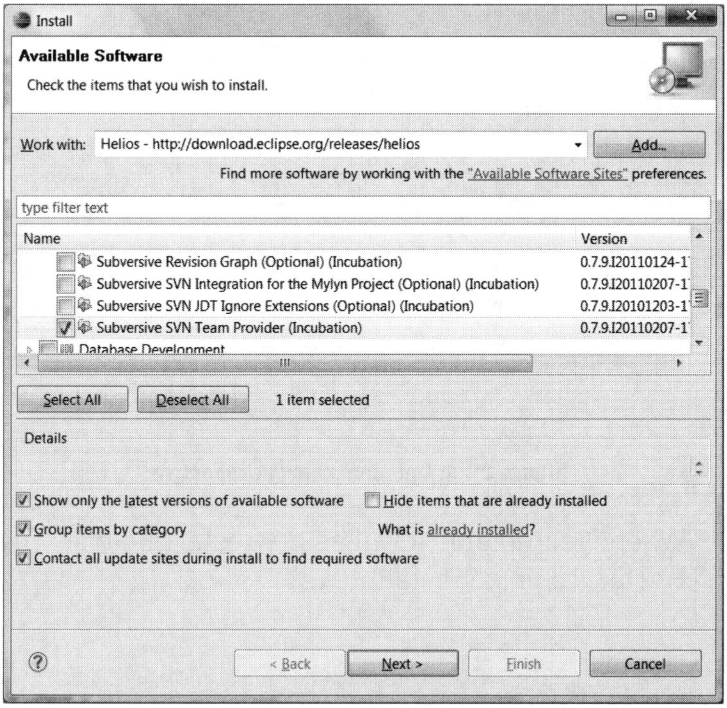

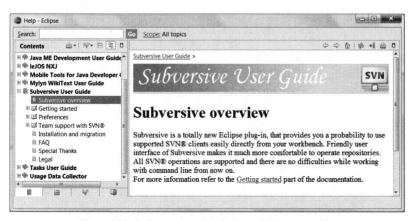

Figure 29-2: Selecting the Subversie plug-in

10.   In the Install window, click Next. The software will calculate requirements and dependencies. When it completes, click Next again. Accept the licensing agreement and continue.

Figure 29-3: Browsing the Subversive user guide

Figure 29-4: Opening a new perspective

11.   When it completes it will ask to restart Eclipse. Click
      Restart to finish installation.

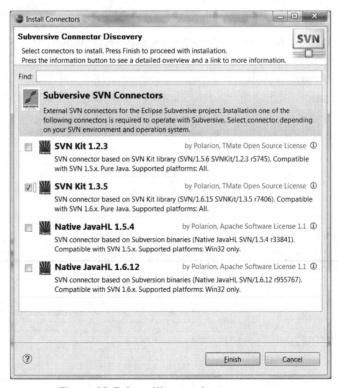

Figure 29-5: Installing one last component

Checking out code

Now that we have Subversive fully installed we can explore the leJOS SVN code repository. Eclipse has different views that change the selection of on-screen components (buttons and panels mainly). The Eclipse team calls these different configurations perspectives. When Eclipse restarts, you will see the SVN Repository Exploring perspective (see Figure 29-6). Now we can import code from the leJOS project.

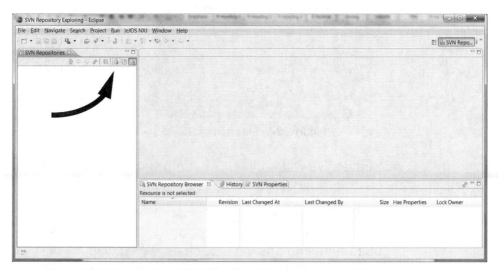

Figure 29-6: Viewing the SVN Repository Exploring perspective

1. On the left side of the Eclipse window is a panel that says SVN Repositories. There are currently none in the list, but we will add the leJOS repository here. Click the New Repository Location button at the top of the panel—it looks like a cylinder with a green plus sign (see Figure 29-6).

2. Where it says URL, enter the following location (see Figure 29-7):

   https://lejos.svn.sourceforge.net/svnroot/lejos

3. Under authentication, enter your Sourceforge user ID and password. Make sure to place a checkmark next to Save authentication (see Figure 29-7). Click Finish.

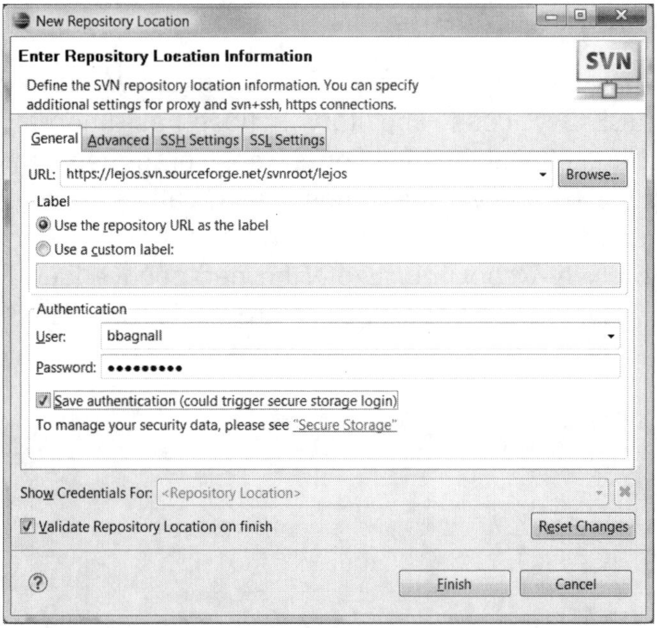

Figure 29-7: Entering the leJOS code location

4. Subversive will contact Sourceforge and add the repository to the SVN Repositories panel.

5. A popup dialog will appear asking if you want to optionally setup password recovery. You can enter a few questions and answers, such as "What was my first pet's name?" If you lose your password, it will ask you these questions to verify your identity.

6. Click the down arrow next to the leJOS project to see the contents. We are interested in trunk. Click the down arrow and you will see a number of sub-projects in leJOS (see Figure 29-8). The one we want to check out is called classes.

7. Right click the classes project, and from the context menu select Check Out. Subversive will download all the classes in this sub-project to your local working directory.

8. Select Window > Open Perspective > Java. You are now back to familiar territory. Now you can see a new project called classes in the left panel. You can browse through the classes, viewing and changing code.

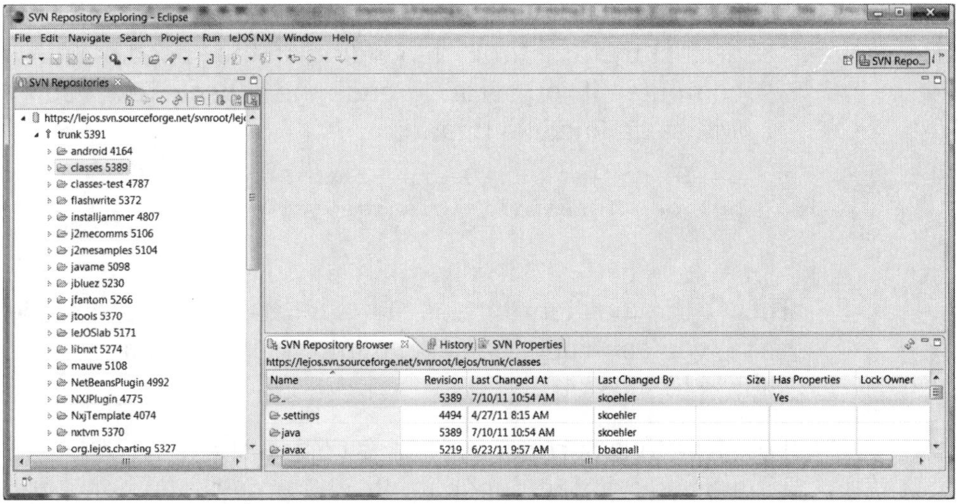

Figure 29-8: Browsing the leJOS subprojects

Building the Project

This section will show you how to modify code in Eclipse and then automatically compile all the classes and package them into a jar file.

1. Let's change one of the class files. We won't modify actual code, just the Javadocs. Open one of the class files, such as lejos.robotics.RangeFinder.

2. The Javadoc class heading starts with /** and ends with */. In between here type something to change the contents of this class file.

3. Pretend you just changed some significant code and want to build the project so you can test how well it works. In the main branch of the classes project there is a file called build.xml (see Figure 29-9). Right click this file and select Run As > Ant Build.

4. The Ant script will compile the classes sub-project and package the classes into classes.jar. It will also build the Javadocs. You can watch the progress in the Console window at the bottom of the Eclipse window.

5. It can take 30 seconds or longer to build the whole project the first time, depending on the speed of your computer. If you scroll through the console text, you will see a line like this:

```
Building jar: C:\Users\Brian\Documents\
helios\classes\lib\classes.jar
```

This is the location the Jar file is located. To use the new code, you can copy this jar file into the lib\nxt directory of your leJOS install.

Feel free to experiment with this copy of the leJOS code. As long as you do not check it into the project you can't hurt the main copy of the code.

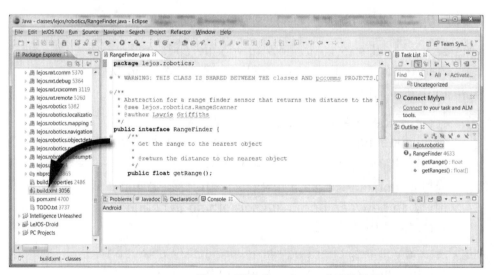

Figure 29-9: Running an Ant Build

Checking in new code

Subversion has a feature to compare your local code against the code in subversion. This helps to identify new code and make sure there are no conflicts if two users have modified the same class at the same time.

1. Select Window > Open Perspective > Other...

2. Select Team Synchronizing. Click OK.

3. We need to add our classes sub-project to the Team Synchronizing list. Click the icon in the upper right of the Team Synchronizing panel (see Figure 29-10).

4. In the Synchronize window select SVN and click Next (see Figure 29-11).

5. Make sure Workspace and Java Workspace are both selected (see Figure 29-12). Click Finish.

6. It will take a few moments to compare the live SVN code with your local copy. When it completes, any new code in SVN will show up in the list with a blue arrow and any outgoing code that you changed will show with a grey arrow (see Figure 29-13).

7. If someone else has checked in new code to SVN, you will see them as updates. To update your code, right click the classes project and select Update.

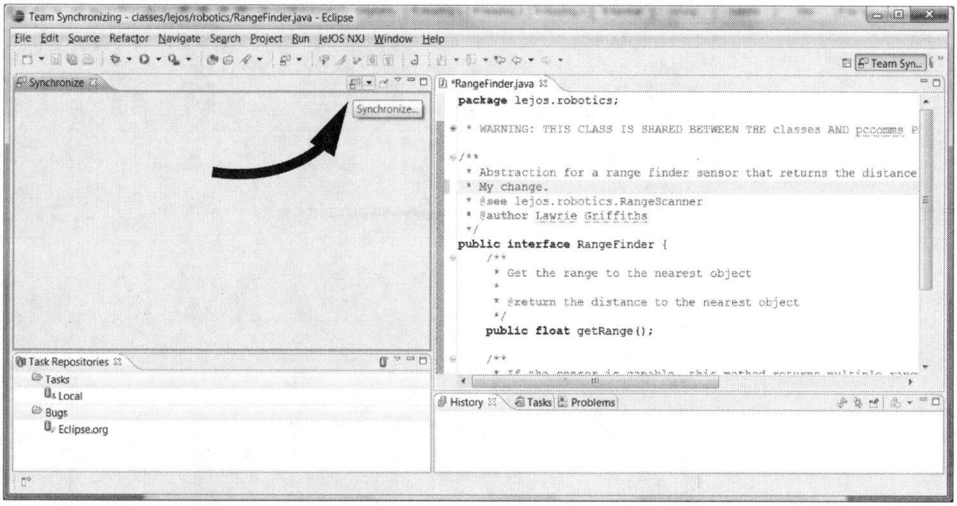

Figure 29-10: Adding a project to team synchronizing

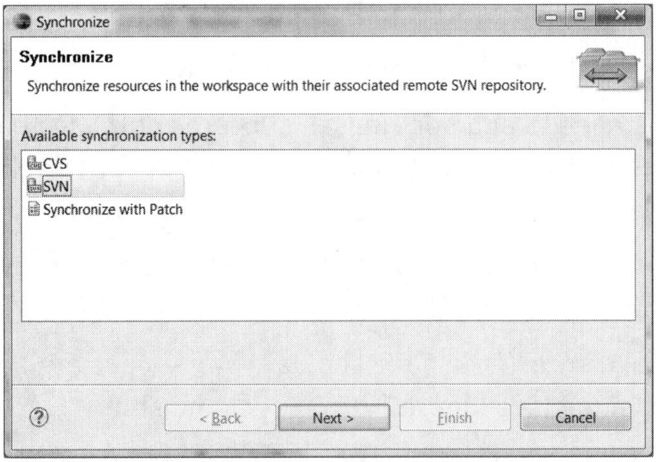

Figure 29-11: Selecting SVN

8. If you become a member of the leJOS project, SVN allows you to commit code. To commit your code, right click either the class you want to check in or the project, and select Commit...

9. Enter a comment for this commit, describing your changes (see Figure 29-14). Click on OK when done and your code will upload to SVN.

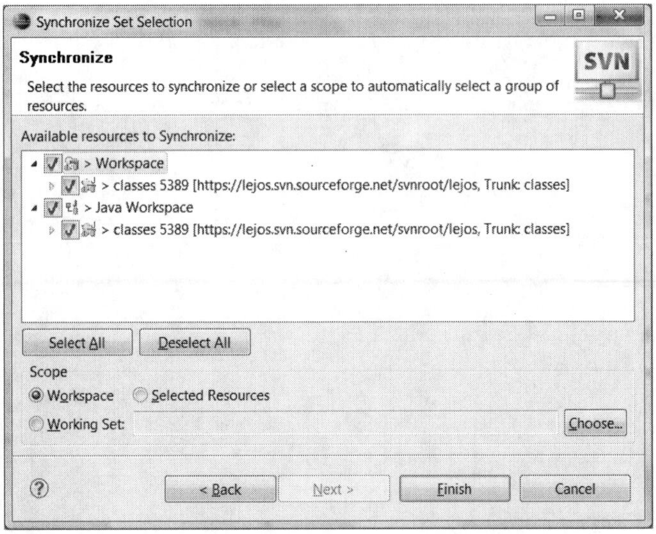

Figure 29-12: Selecting the workspace

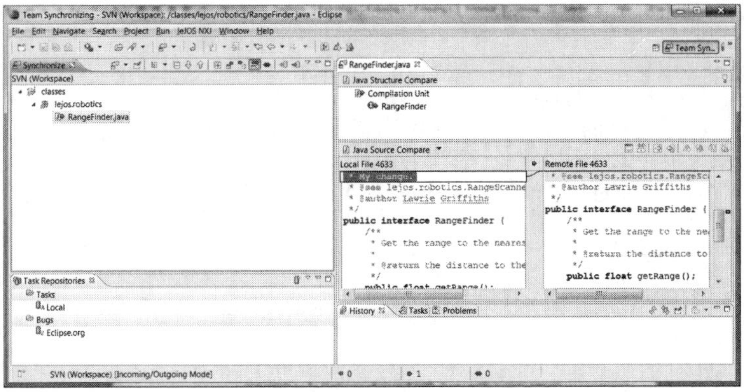

Figure 29-13: Comparing code

If you tried to go through with step 9, it will not allow you to upload the code because you do not have the proper rights. If you want to commit code, one of the leJOS project admins needs to add your name to the leJOS project. Visit the leJOS project page at Sourceforge and email one of the administrators. They will probably want to know what you have in mind for the project, so be prepared to tell them what class you have made changes to or what you are interested in developing.

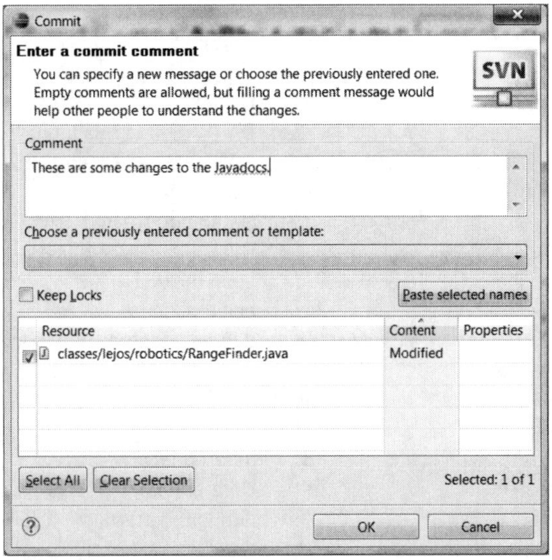

Figure 29-14: Committing a change

WITHDRAWN

24.95   2/21/12.

LONGWOOD PUBLIC LIBRARY
800 Middle Country Road
Middle Island, NY 11953
(631) 924-6400
mylpl.net

LIBRARY HOURS

Monday-Friday	9:30 a.m. - 9:00 p.m.
Saturday	9:30 a.m. - 5:00 p.m.
Sunday (Sept-June)	1:00 p.m. - 5:00 p.m.